INSECT HORMONES

UNIVERSITY REVIEWS IN BIOLOGY

General Editor: J. E. TREHERNE

Advisory Editors: Sir VINCENT WIGGLESWORTH, F.R.S.
                    M. J. WELLS
                    T. WEISS-FOGH

PLATE I

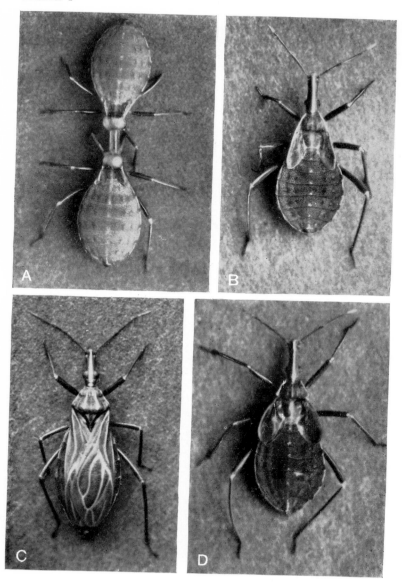

PLATE I. A. 4th-stage larva of *Rhodnius* decapitated after the critical period, connected by a glass tube to larva decapitated at one day after feeding. B. Normal 5th-stage larva of *Rhodnius*. C. Normal adult. D. Giant or 6th-stage larva produced by implantation of an active corpus allatum into a 5th-stage larva.

INSECT HORMONES

V. B. WIGGLESWORTH

Department of Zoology
University of Cambridge

W. H. FREEMAN AND COMPANY
SAN FRANCISCO

BY THE SAME AUTHOR:

Insect Physiology
(Methuen, 1934; 6th ed. 1966)

The Principles of Insect Physiology
(Methuen, 1939; 6th ed. 1965)

The Physiology of Insect Metamorphosis
(Cambridge University Press, 1954)

The Control of Growth and Form
(Cornell University Press, 1959)

The Life of Insects
(Weidenfeld & Nicholson, 1964)

OLIVER & BOYD
Tweeddale Court
Edinburgh EH1 1YL
A Division of Longman Group Ltd
Library of Congress Catalog Card Number: 74 134310
International Standard Book Number: 0 7167 0688 1

First Published 1970
© 1970 Sir Vincent Wigglesworth
All rights reserved

Printed in Great Britain by
R. & R. CLARK, LTD., EDINBURGH

Preface

The study of insect hormones has grown up since 1917 and most of our knowledge has been gained in the last thirty years. At the present time the literature of the subject is expanding at a very rapid rate; and this makes it only too easy to lose sight of the main outlines in a mass of disputes about detail.

The purpose of this introduction to the study of insect hormones is to concentrate on the main themes that are well established, and to illustrate them by concrete examples which can be reasonably well documented. The treatment must therefore be somewhat arbitrary; and the student who wishes to make a more complete survey must follow up the references given.

At the same time the subject matter of 'insect hormones' is interpreted in the widest possible manner. Besides the chemical substances which are liberated from regular endocrine glands and circulate in the blood to influence the growth and metabolism of remote parts of the body, I have included in the later chapters some reference to 'neurohumours', 'tissue hormones' and 'inductors', 'gene hormones' and 'pheromones'—materials which it can be argued are not properly called hormones at all.

V. B. Wigglesworth

Contents

1: Hormones, Growth and Moulting

The earliest discovery of a glandular secretion into the blood of animals was made by Berthold in the middle of the last century, when he proved that the masculinizing effect of the testis in birds is due to a substance liberated into the circulation. This discovery led a number of authors, during the later years of the nineteenth century, to study the sexual glands of insects in order to test whether they are the source of a similar hormone. Castration in both sexes, with or without the implantation of gonads from the opposite sex, was carried out in a variety of insects: the silkworm *Bombyx*, the gypsy moth *Lymantria*, the cricket *Gryllus*, among others. All these experiments gave negative results and it became generally accepted that not only did the gonads of insects not secrete hormones, but (and this was an unjustifiable conclusion) that insects did not secrete any hormones at all.

The 'moulting hormone'

This dogma delayed research and it was not until 1917 that Kopeč discovered that if full-grown caterpillars of *Lymantria* were ligated halfway along the body, the anterior half duly pupated but the posterior half remained unchanged as a larva; and that removal of the brain from such full-grown caterpillars likewise prevented pupation—although starved larvae from the same batch, with the brain intact, pupated and developed into adult moths. If the nerve cord was cut through just behind the head, pupation and further development were not affected. From these experiments Kopeč concluded that the brain was the source of a hormone necessary for growth and metamorphosis.

The suggestion that the brain was a secretory organ was surprising; it prejudiced the acceptance of these results. When Hachlow in 1931 experimented with the pupae of various butterflies (*Gonepteryx*, *Vanessa*, etc.), which he transected at different levels and then sealed

1

the cut ends by fixing them with wax to glass plates, he found that it
was always the fragment that included the thorax which alone completed
its development to the adult.   From this he concluded that some
centre of an undetermined nature (he did not specify a hormone source)
situated in the thorax, was controlling growth and metamorphosis.[212]
Experiments on the same lines as those of Kopeč were carried out in the
blowfly *Calliphora* in 1935 by Fraenkal[212] who likewise found that if the
larvae, a few hours before puparium formation was due to occur, were
ligatured around the middle, the cuticle of the anterior half was trans-

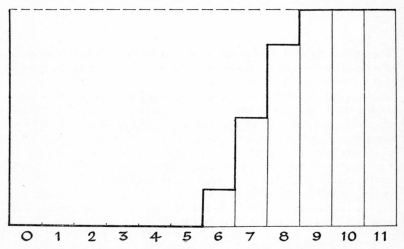

FIG. 1.   Effect of decapitation on moulting of *Rhodnius* 5th-stage larva.
Ordinate: proportion of larvae moulting; abscissa: number of days after
feeding when a given batch was decapitated.

formed into a hard and dark 'puparium'; the posterior half retained
the white and soft character of the larva.

At about this same time similar results were obtained in the blood-
sucking bug *Rhodnius*.[211]   This insect has five larval stages (Plate IB)
in each of which it takes only a single gigantic meal of blood; it then
moults to the next instar, ten to twenty days later according to which
larval stage is being studied.   Thus the 4th-stage larva requires about 14
days between feeding and moulting (at 26°C).   If the head is removed and
the neck sealed with paraffin wax soon after feeding, no growth occurs
and the larva does not moult—although such larvae have remained alive
for many months.   But at about four days after feeding (in the 4th stage)
there is a 'critical period', such as had been observed by Kopeč in
Lymantria, and after this critical period moulting takes place normally

even after decapitation (Fig. 1). The headless insect very rarely sheds the old cuticle; but otherwise the entire process of growth and cuticle formation, often with normal hardening and pigmentation, is completed.

Here again there was clear-cut evidence that the moulting process was initiated by a hormone present in the circulating blood; for if two larvae of *Rhodnius* were decapitated, one at twenty-four hours after feeding (that is, before the critical period) and the other at seven or eight days after feeding (that is, after the critical period) and these two insects were then joined in parabiosis by applying the two cut ends together and sealing the join with paraffin wax—then the twenty-four hour insect also would be induced to moult. The same result was obtained if the cut ends were not brought directly into contact but were connected by a glass capillary several millimetres long (Plate IA).

Source of the ' moulting hormone '

During the 1930s it was discovered by Ernst Scharrer that the brain in many animals contains cells which often give the histological appearance of secreting cells. They were called 'neurosecretory cells' and were found in mammals and other vertebrates, and in many invertebrates, including the honey-bee among insects. Hanström undertook to look for cells of this type in *Rhodnius* and he readily demonstrated their presence in the dorsum of the brain.[212] It was then found that if the region containing the neurosecretory cells was cut out from the brain of a *Rhodnius* larva a few days after feeding, and was implanted into the abdomen of a larva decapitated before the critical period, this larva could be induced to moult.[211] No other part of the brain had this effect. This was indeed the first demonstration of a function (and an endocrine function) for the neurosecretory cells in any animal.

But it will be recalled that Hachlow had shown that in the pupa of Lepidoptera the centre necessary for growth and moulting appeared to be located in the thorax. And it was found by Burtt[22] in *Calliphora* and by Hadorn and Neel[212] in *Drosophila* that if the ' ring gland ' of Weismann (Fig. 2) was excised the larva was no longer able to form the puparium; and that ' ring glands ' taken from mature *Drosophila* larvae and implanted into younger larvae, will induce puparium formation before the normal time. It appeared that in these insects this thoracic gland was the source of the hormone.

As long ago as 1762, in his classic work on the goat moth caterpillar *Cossus*, Lyonet had described what he called ' granulated vessels ' in the fore-part of the thorax. It was found by Fukuda[212] that it is this gland, commonly called today the ' prothoracic gland ', which is the source of the moulting hormone in the silkworm *Bombyx*.   If the larva of the silk-worm is ligatured behind the prothoracic gland, only the anterior half pupates.   But the posterior part can be induced to pupate if the pro-thoracic gland is implanted into the abdomen.   At the critical period the prothoracic gland releases an active principle into the blood. This same gland controls the moulting of the larva and the development of the pupa.   The isolated pupal abdomen will resume its development if it is connected to the anterior part of the body by a capillary tube, or if the prothoracic gland removed from another pupa or from a larva is implanted into it.

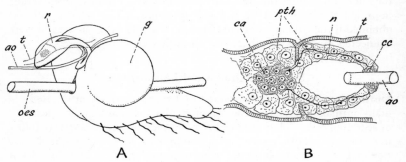

A   B

FIG. 2.   A, Diagram to show the relation of the ring gland in *Drosophilia* to the central nervous system and the aorta.   B, diagram showing details of ring gland.   *ao*, aorta; *ca*, corpus allatum; *cc*, corpus cardiacum; *g*, fused brain and ventral ganglia; *n*, nerve axons from corpus cardiacum to corpus allatum; *oes*, oesophagus; *pth*, prothoracic gland cells; *r*, ring gland; *t*, tracheae.   (After King, Aggarwal & Bodenstein).

There was thus clear evidence that both the brain and the prothoracic gland are concerned in the control of growth and moulting in the Lepidoptera.   It was therefore suggested by Piepho[212] in 1942 that the secretion from the prothoracic gland may be produced in response to the secretion from the brain—just as the production of thyroxine by the thyroid gland is secondary to the action of the thyrotropic hormone of the pituitary; and that in Lepidoptera (as had, indeed, been suggested in *Rhodnius*) not a single active principle but a succession of active principles is responsible for the moulting process.

It had been noted by Plagge in 1938[212] that although the implanted brain would cause pupation in *Celerio* larvae from which the brain had been removed, it would *not* induce pupation in the posterior fragments of ligated larvae.   The significance of this was proved experimentally by Williams[212] who repeated and extended this experiment on the pupa of the American silk moth *Hyalophora cecropia*.   An active brain implanted in the isolated abdomen of the pupa will not induce development; whereas it will cause development in the anterior half from which the brain has been removed.   On the other hand, the isolated abdomen will develop if it is provided with an active brain plus prothoracic glands. As in *Rhodnius*, the actual source of the brain secretion is the neuro-secretory cells.

This two-stage process of hormone secretion appears to be general in insects.   In *Calliphora*, as studied by Possompès, the ' ring gland ' is the immediate source of the hormone necessary for puparium formation, but it needs to be activated by the neurosecretory cells in the brain.[216]   In *Rhodnius*, also, a ' thoracic gland ' has been discovered. As in *Hyalophora*, once it is activated it will induce moulting when implanted into a larva decapitated before the critical period, or into the isolated abdomen of such a larva.   The active brain, on the other hand, is effective in the decapitated larva (which retains its ' thoracic gland '); it is ineffective in the isolated abdomen (Fig. 3).[212]

It is customary to refer to the brain hormone as the ' *activation hormone* '; and the prothoracic gland hormone as the ' *moulting hormone*', the active chemical responsible being named ' ecdysone '.

The endocrine organs concerned in growth

Cerebral and retrocerebral organs

The cerebral organs comprise the *neurosecretory cells* of the pars intercerebralis in the dorsum of the brain.   These may consist of a single group of five or six large cells on each side, as in the milkweed bug *Oncopeltus*, or they may consist of large numbers, 1,000–2,000 of much smaller cells, as in the locusts.   They are classified in some four or more types characterized by their staining properties (p. 89).   The most characteristic type elaborates an acidophil product staining with acid fuchsin or with phloxin, which becomes strongly basophil after per-manganate oxidation, and then stains deeply with the chromehaema-toxylin of Gomori.   It stains strongly also with the paraldehyde fuchsin of Gomori.   Another characteristic property of the substance of the neurosecretory cells is the luminous blue appearance (Tyndall's blue) which they show when examined under dark-ground illumination (p. 89).

The neurosecretory cells of the pars intercerebralis commonly form two groups, a medial and a lateral group—as in *Hyalophora*, and in locusts, cockroaches, etc.   In *Hyalophora* it is said that these cells are effective in inducing development only if the secretion from the two groups is mixed and combined.[216]   Like other neurons the neuro-secretory cells have axons; and the characteristically staining contents of the cell body pass down these axons and may sometimes be revealed by appropriate staining methods or by dark-ground illumination. The axons from the medial group cross over in the midline of the brain, leave the back of the brain by way of the medial corpus cardiacum nerves and run to the corpus cardiacum. The axons from the lateral group do not cross over but form the lateral corpus cardiacum nerves to supply the corpus cardiacum of the same side (Fig. 26).

The *corpus cardiacum*, in which most of the neurosecretory axons terminate, is also nervous in origin. It is a part of the stomatogastric nervous system (see p. 90) and like the remainder of that system it arises

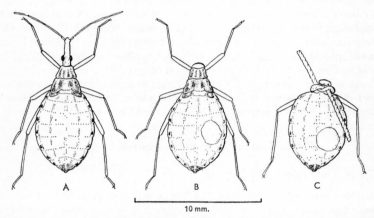

10 mm.

FIG. 3.   A, normal 4th-stage larva of *Rhodnius*; B, the same decapitated with implant in abdomen; C, the same ligatured through metathorax with implant in the isolated abdomen.

from an ingrowth of the dorsal wall of the stomadoeum. It has a most curious structure. Much of it is made up of the bulbous endings of the neurosecretory axons from the brain; but it also contains the cell bodies of neurosecretory cells which produce a secretion of their own (p. 91). Those parts of the corpus cardiacum that are laden with the product of the neurosecretory cells of the brain show the same staining reactions and the same luminous appearance under dark-ground illumination. The corpus cardiacum, in fact, seems to serve as a storage organ for this secretion—presumably releasing it or some modified product of it into the circulating blood as required. If the nerves bringing this material are cut or ligatured, the region of the nerve above the obstruction becomes swollen with the neurosecretory product. The corpus cardiacum is in origin a median structure, but commonly appears in the form of two lobes connected by a bridge; and sometimes, as in mosquitoes, the two halves may be quite separate.

It is commonly believed that the ' activation hormone ' derived from the neurosecretory cells passes through the connective tissue sheath around the corpus cardiacum in a dispersed form and is carried around the body in the circulating haemolymph.[216] It was found by Johannson[216] that extirpation of the ten neurosecretory cells of the pars intercerebralis in *Oncopeltus* did not prevent moulting. He assumed that the already formed neurosecretory material, which in this insect is stored in the wall of the aorta rather than in the corpus cardiacum, was sufficient for this purpose.

The neurosecretory cells in some insects show distinct cycles of secretion during the most active periods of growth. In the meal moth *Ephestia*, as studied by Rehm, there is an intense discharge of material from the cell bodies during the critical period of the prepupal stage, and again during early pupal development; and at the time when the material is disappearing from the neurosecretory cells the corpus cardiacum shows a large increase in volume.[216] Likewise in *Bombyx* there is a similar increase in volume after each burst of secretion from the neurosecretory cells.[216]

On the other hand, although the experimental evidence proves that the neurosecretory cells in *Rhodnius* are liberating their product very soon after the ingestion of the blood meal, it has not been possible to observe any distinct changes either in the cells themselves or in the corpus cardiacum at this time. It may well be that more careful study will demonstrate such changes. It is generally agreed since the work of Rehm that in the Lepidoptera, *Ephestia*, *Galleria* and *Pieris* the most striking accumulation of stainable material in the cytoplasm of the neurosecretory cells takes place *after* the period of active liberation of the secretion[216] (see pp. 82, 93).

Closely associated with the corpus cardiacum is the *corpus allatum*, which will be considered in later chapters (pp. 45, 71). The corpora allata were the first organs in insects to be recognized as possible ductless glands. They arise by budding of ectodermal cells between the mandibular and maxillary segments. Later these cell nests become separated from the epidermis and form compact deeply staining bodies. In the primitive apterygote insects (*Lepisma*, etc.) and in the Phasmida (stick insects, etc.) each consists of a hollow vesicle covered by a single layer of columnar cells. Like the corpora cardiaca, in Hemiptera and in the higher Diptera they fuse to form a single median structure. They are innervated from the nerves to the corpora cardiaca, and the neurosecretory product can often be seen extending into these nerves also. Precisely what the function of this product may be is not known.

Prothoracic gland system

The origin of the prothoracic glands was described by Toyama in the silkworm at the very beginning of this century long before their function was known. They arise as ingrowths from the ectoderm of the second maxilla, in close association with the salivary glands. They have the same origin in *Rhodnius* and other Hemiptera, and are carried backwards into the thorax along with the salivary glands. In the more primitive groups of insects, Thysanura, Ephemeroptera, Odonata, Isoptera, Dermaptera, Acrididae, they remain as compact bodies in the head and are named ' ventral glands '. In the stick insect *Carausius*, they occur in two parts: a ventral gland in the head and a so-called ' pericardial gland ' in the thorax.[216] In other insects, such as Blattidae, Hemiptera, Coleoptera, Lepidoptera and Hymenoptera the components form loose chains of cells or lace-like meshworks. In the larvae of primitive Diptera (the Nematocera) the gland is always associated with conspicuous tracheae and is sometimes called the ' peritracheal gland '; but in the higher Diptera the ' peritracheal glands ' unite dorsally with a median fused corpus allatum, and ventrally with the corpus cardiacum to form Weismann's ' ring gland ' which surrounds the aorta as this passes above the brain. The ' ring gland ' is thus a composite structure: it is the two side arms, made up of large cells, which represent the ' prothoracic gland ' (Fig. 2). In order to have a single name for all these diverse glands it has been suggested[74] that they should be called ' ecdysial glands '. This term may gain acceptance; unfortunately it has been applied for many years to the dermal glands which are believed to secrete the ' cement layer ' over the cuticle at the time of ecdysis.

The prothoracic glands are usually well supplied with tracheae, and this is an aid to finding them in dissection. They are richly supplied with nerves in Lepidoptera, in cockroaches and in grasshoppers; in *Rhodnius* and in some other Hemiptera no nerve supply has been found.

The prothoracic gland cells have a fine structure which resembles that of steroid secreting cells in vertebrates (such as the interstitial cells of the testis in mammals). In Blattids, for example, the cells are covered by interlacing processes; their cytoplasm contains only scanty granular endoplasmic reticulum; there is no participation of Golgi elements in formation of the secretory product and no obvious storage of secretion.[74, 174] The cells undergo a conspicuous cycle of secretory activity during each instar, when exposed to the activation hormone from the brain. In *Rhodnius*, for example, the nuclei become greatly enlarged and lobulated during the ' critical period ' when moulting is beginning; the cytoplasm also becomes more extensive and more basophil. Later in the moulting cycle, and well before the old cuticle is shed, the cells revert to their resting condition.[212]

A cycle of this kind takes place in each larval stage; but when the insect undergoes metamorphosis to the adult form the cells of the prothoracic gland degenerate and disappear so that the adult insect cannot continue to moult (Fig. 4). In Odonata degeneration begins within 3 hours after the imaginal moult; in the earwig *Forficula* and in *Rhodnius* autolysis is well advanced in 24 hours and complete in 2 days; in *Periplaneta* the gland breaks down more slowly but it has usually gone by the fourteenth day after metamorphosis.[212] In the primitive

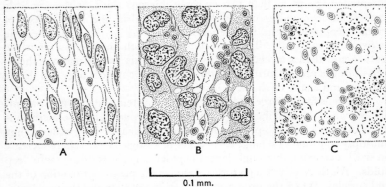

0.1 mm.

Fig. 4. Changes in the cells of the prothoracic gland during moulting in the 5th-stage larva of *Rhodnius*. A, unfed larva with inactive gland cells. B, at 10 days after feeding with cells at peak of activity. C, at one day after moulting to adult stage; gland nuclei have broken down.

Thysanura, *Thermobia*, *Lepisma*, etc., which alternately moult and reproduce throughout their adult life, the prothoracic glands (that is, the ' ventral glands ') persist.[207]

The *raison d'être* of the moulting hormone: the moulting process

The epidermis is the chief tissue responsible for growth and form in the insect. At each instar the visible form of the body is defined by the cuticle which is laid down by and is indeed an integral part of the epidermis. In the softer parts of the integument the cuticle can unfold and stretch, but in the more rigid skeletal structures, such as the head and the appendages, growth is impossible unless the cuticle is shed.

It follows, therefore, that growth and moulting are intimately associated. During the greater part of the life of the insect the epidermal cells are firmly attached to the cuticle; they are engaged in secreting or in maintaining its substance.   There is only a brief period, when the cells are detached from the cuticle, during which they can multiply and the epidermis can grow.   Cuticle formation and cellular growth are mutually incompatible.   Growth in the insect, therefore, of necessity takes place in cycles—periods of growth alternating with periods of cuticle formation.   Hence it follows that the hormone that initiates the moulting process is the hormone responsible for growth.

The cycle is seen most clearly in an insect such as *Rhodnius* which is normally subject to more or less prolonged periods of fasting during which the epidermal cells are in a state of dormancy.   They are attenuated, no more than two or three microns thick, with very little cytoplasm, containing only a few mitochondria and only traces of ribonucleoprotein (Fig. 6).   When moulting begins these cells become active and enlarged.   They presumably detach themselves from the cuticle and then proceed to multiply by mitosis and to arrange themselves in the proper form to lay down the new cuticle which will define the size and shape of the next instar.

The cells first enlarge and build up their protein and enzyme content. They then divide and increase in number—more actively in some regions than in others, so as to provide for changes in shape.   The cells as they multiply continue to arrange themselves in a single layer.   Where increase in size is occurring this single layer of epidermal cells is thrown into deep folds.   All these changes go forward under the passive protection of the old cuticle.   When the process of growth is complete and the new instar (often in highly convoluted form) has been fully organized and mitosis is at an end, the deposition of the new cuticle begins (Fig. 5).

First a boundary membrane, or outer epicuticle, which appears in electron microscope sections as a dense triple layer about 150 Å thick, is laid down over the surface of the epidermal cells.   This is followed by a more substantial layer of lipoprotein, the inner epicuticle.   Then the main substance of the cuticle composed of chitin and protein is laid down in a succession of laminae.   Those regions of the cuticle that are destined to be hard in the next instar become impregnated with more protein that will later be subject to tanning.   Then the epidermal cells secrete the enzymes chitinase and proteinase into the space between the old cuticle and the new.   This ' moulting fluid ' acts upon the inner layers of the old cuticle and digests them and the products of digestion are absorbed through the new cuticle by the epidermal cells.   Finally nothing remains of the old cuticle but the indigestible ' epicuticle ' and the tanned or ' sclerotized ' layers that form the hard ' exocuticle '.

The remnants of the old cuticle still retain their waterproofing properties.   Before this covering is shed the new cuticle must be waterproofed.   The new cuticle is deposited around canals, the so-called

' pore-canals ' that extend from the epidermal cell to the inner limit of the epicuticle.   Within the pore canals are finer filaments (the wax canal filaments) which run from the cells and pierce the epicuticular layers to open freely at the surface of the cuticle.   These filaments are only about 100 Å thick and are apparently concerned in the discharge of the water-proofing wax upon the cuticle surface.   When this is laid down the

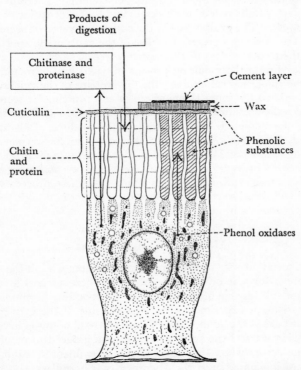

FIG. 5.   Schematic figure summarizing the activities of the epidermal cell during cuticle formation.

insect is waterproof.   The final act of moulting or ' ecdysis ' now takes place.   The insect begins to swallow air to increase its volume.   By contraction of the intersegmental muscles in the abdomen the body fluid is driven forward to expand the soft cuticle of the head and tho-rax, so that the old cuticle ruptures along preformed lines of weakness where an exocuticle is absent, and the insect slowly withdraws itself from the old skin.

Soon after moulting, in some insects even before the old skin is shed, a further layer, the so-called ' cement layer ', is poured out by dermal glands whose ducts open through the cuticle, and spreads over the freely exposed wax and quickly hardens to form a protection for it.   The new cuticle of the insect is still soft and colourless.   During the next few hours phenolic derivatives of the amino acid tyrosine already secreted into the cuticle are oxidized to form quinones which tan and harden the protein of the exocuticle to form the amber-coloured ' sclerotin '; and in certain parts similar oxidative processes lead to the formation of melanin in the cuticle.   The greater part of the cuticle has yet to be secreted; the bulk of the chitin and protein lamellae of the endocuticle are laid down during the next few days or the next few weeks.

Thus, during the moulting process initiated by the moulting hormone, the epidermal cell secretes the lipoproteins of the epicuticle, and then lays down the inner layers of chitin and protein.   It secretes and discharges the enzymes proteinase and chitinase which digest the inner layers of the old cuticle, and it reabsorbs the products of digestion.   It produces the phenolic compounds that will later be responsible for hardening and darkening, and then the long-chain waxes to waterproof the surface.   Finally it produces the phenol oxidase and other enzymes which lead to melanization and tanning.   All these processes are integrated and timed so that they follow one another in orderly sequence, and synchronously in all parts of the body.

Moulting of adult insects

We have seen that very soon after the insect becomes adult the thoracic glands break down and disappear.   Moulting therefore ceases.   But the tissues are still capable of responding to the moulting hormone if they are exposed to it experimentally.   The antenna of the earwig *Anisolabis*[212] or tarsi of the adult *Gryllus*[212] transplanted onto the integument of a growing larva of the same species can be induced to moult again. The same occurs when fragments of the integument of adult Lepidoptera are implanted into the abdomen of caterpillars.[212]   And the adult bed-bug *Cimex* can be made to moult by joining it to a moulting larva of *Rhodnius*.[211]

The new cuticle laid down by the adult *Cimex* when it moults again is an amorphous structure with no properly formed bristles, if an adult several weeks old is used.   But if a young adult *Cimex*, or an adult *Rhodnius* of any age, is caused to moult by joining it to a moulting fifth-stage larva, or by implanting into the abdomen the prothoracic glands taken from a fifth-stage larva around the critical period,[212] it lays down a cuticle in which the pattern and structure and the form of the bristles are indistinguishable from those of a normal adult.

The adult silkmoth *Hyalophora* can be made to moult by joining it in parabiosis with a developing pupa—but, like the adult *Cimex* several weeks old, it lays down a cuticle devoid of hairs or scales.   The adult insect does not have the necessary ' ecdysial lines ' where the old cuticle can easily be broken and shed when moulting takes place; it remains enclosed within the old skin.   But it has been possible to induce the adult *Rhodnius* to make two such moults.   Whereas the adult will react in this way to implantation of the prothoracic glands, it fails to respond to implantation of the neurosecretory cells from the brain, since it has no prothoracic glands to be activated.

Stimuli to secretion of the ' activation hormone '

In many insects one cycle of growth and moulting succeeds another without noticeable pause.   In these it is not easy to recognize the nature of the stimuli which bring about the secretion of the hormones concerned.   What little information exists on this subject has been obtained in insects with discontinuous growth, or insects with well-defined periods of arrested growth or diapause.

It was early shown that in *Rhodnius*, which takes large and infrequent meals of blood, it is the degree of distension of the abdomen which produces the stimulus to the liberation of the ' activation hormone ' from the brain.   The state of nutrition is not responsible—for repeated small meals of blood (each of them as great as the insect's own weight) which keep the gut continuously supplied with undigested blood, will not initiate moulting.   A large meal is necessary.

This degree of distension provides a nervous stimulus to the brain. If the conduction of this stimulus is interrupted by section of the nerve cord in the neck, moulting does not begin.[211]   Van der Kloot[216] has demonstrated the existence of stretch receptors in the abdomen of *Rhodnius* (one on each side of each abdominal segment) which lead to the appearance of impulses in the nerves to the corpus cardiacum.   These receptors ' adapt ' hardly at all: they continue to discharge as long as the abdomen is expanded.

In *Locusta migratoria* in which feeding is more or less continuous and one moult follows rapidly upon another, Clarke and Langley[36] suggested that stretch receptors in the wall of the pharynx are stimulated during the act of feeding.   These sense organs are innervated from the frontal ganglion which connects up with the brain (Fig. 25).   Excision of the frontal ganglion prevents moulting.   But it has long been known that the frontal ganglion is necessary for swallowing: the arrest of moulting after its removal may be the result of starvation (see p. 81).

There are many examples in other insects of growth being influenced by stimuli which act presumably upon the brain. We shall discuss some of these in connection with diapause (p. 29); other examples may be quoted here. The larva of the wheat stem sawfly *Cephus cinctus* is reluctant to begin post-diapause development once it has been removed from the stub of the wheat straw; hormone secretion is influenced by the 'feel' of the surroundings.[216] Likewise in the larva of the honey-bee, contact with the silk cocoon (or with a similar cell of gauze) provides a necessary stimulus to the brain for the initiation of pupation.[216] And it was shown by Mellanby[216] that larvae of the blowfly *Lucilia* which are in a state of diapause in dry sand can be caused to pupate simply by placing them in empty glass vials out of contact with sand. Under these conditions the ring gland shows histological signs of renewed activity, but the stimulus has presumably come from the peripheral sense organs via the brain. Similarly in the fleshfly *Sarcophaga perigrina*, pupation is delayed by contact with water; but ecdysone secretion is restored within 6 hours after transference to dry surroundings, and the larvae pupate 18–24 hours later.[149] Experimental distortion of the body may likewise inhibit moulting in caterpillars of *Galleria*.[49]

Detectable effects of the 'activation hormone'

In the larva of *Calliphora* the brain exerts its action on the prothoracic gland (that is, the side arms of the ' ring gland ' of Weismann) only when the nervous connection between the brain and the ring gland is intact.[212] In other insects the hormone reaches the prothoracic gland via the circulating blood. But perhaps this difference is not very great; for in *Calliphora* the corpus cardiacum is an integral part of the ring gland and the transfer from the brain to the corpus cardiacum in all insects is by way of the nerves. According to Fraser[216] the active material, after reaching the corpus cardiacum of *Calliphora* via the nerves, is then liberated into the blood, as in other insects.

The nature of the action of the brain hormone on the prothoracic gland has been little studied. It is no brief ' triggering ' action; the prothoracic gland must be exposed to the hormone continuously until the moulting process is well established. In *Rhodnius* the first signs of renewed growth within the epidermis are apparent within a few hours after feeding. But the process stops and the cells revert to the resting state if the head is removed before about 4 days after feeding in the fourth-stage larva, before about 7 days in the fifth-stage larva.[211] These times are what is commonly referred to as the ' critical period '. Likewise in the larva of *Cephus*, the brain continues to influence the prothoracic glands even after they have begun to secrete; and if the production of the brain hormone is halted by high temperature (35°C) the larva reverts to the resting state.[216]   On the other hand, in the larva of the wild silk moth *Platysamia cynthia*, the prothoracic gland seems to release the pupation hormone continuously once it has been stimulated by the brain hormone.[216]

In the pupa of *Antheraea*, within 12 hours of exposure to the ' activation hormone ' the nuclei of the prothoracic gland are taking up labelled uridine and incorporating it into the RNA of nuclei and cytoplasm; and this is followed by the uptake of tritiated amino acids as protein synthesis begins.[145]   But this tells us little about what is happening. The suggestion has been made that the neurosecretory product from the brain may perhaps provide the raw material from which hormones manufactured by different parts of the endocrine system may be derived, and that in the prothoracic gland it may be converted into the moulting hormone.[216]   At present there is no solid experimental evidence to support these ideas.   But we have already noted that there is histological evidence that the neurosecretory product from the brain finds its way along the axons to the corpus allatum also.   In the larva of the Colorado potato beetle *Leptinotarsa* the corpora allata increase in volume before the final moult, and fine droplets of neurosecretion can be seen in the nerves ramifying within them.   The same is to be seen in the larva of *Bombyx* and of Odonata.[216] And a study of the corpus allatum of the Madeira cockroach *Leucophaea* with the electron microscope has revealed the axon terminals from the neurosecretory cells with only a thin membrane separating the neurosecretory granules from the corpus allatum cells.[216]

These observations are of interest because it has been found both in *Hyalophora* and in *Bombyx* that the implantation of corpora allata will sometimes induce a pupa deprived of the brain to resume its development—as though it were providing a source of the activation hormone.

Detectable effects of the prothoracic gland hormone (ecdysone)

The effects of the moulting hormone are most clearly seen in insects with a dormant period between successive moults—as in the bloodsucking bug *Rhodnius*, or in insects with a definite period of diapause, or arrested growth.   In the abdomen of *Rhodnius*, for example, there are two tissues which are specially concerned in the processes of growth and moulting.   First and foremost the epidermis, which will lay down the new and larger cuticle as described above.   Secondly there are the intersegmental muscles.   These have no function in the ordinary life of the *Rhodnius* larva: they come into operation only during the act of moulting when they contract and increase the pressure on the haemolymph and thus drive the body fluid forward into the head and thorax to rupture the old cuticle, and when the insect has cast the old skin the same increase in blood pressure is available to expand the legs and wings before they harden.   As soon as moulting and hardening are complete the contractile fibres of the intersegmental muscles undergo autolysis and the proteins are utilized elsewhere in the body (Fig. 34).   Almost nothing remains but the nuclei and the muscle sheath.

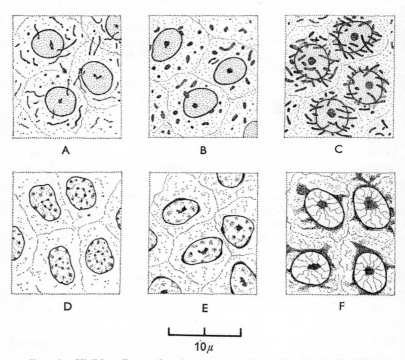

FIG. 6.   Visible effects of ecdysone on epidermal cells of *Rhodnius*. Upper row shows changes in nucleoli and mitochondria: A, newly fed insect; B, after one day; C, after four days.   Lower row shows changes in nucleic acids: D, newly fed insect; E, after one day; F, after 4 days.

In the dormant state between moults the cells in the epidermis and in the intersegmental muscles are inactive; the nuclei have very small nucleoli; the cytoplasm contains only traces of RNA and a few small granular or filamentous mitochondria (Fig. 6).   They are aroused from this state by the moulting hormone. The process is commonly called 'activation'.   Within three or four hours after feeding the nucleoli begin to enlarge; ribonucleoprotein appears in the cytoplasm and rapidly increases in amount.   The mitochondria enlarge and then break up and multiply.   Protein synthesis becomes active; the cytoplasm increases in amount, and in the muscles the new contractile fibrils are formed. Meanwhile the DNA in the nuclei increases and this leads on to mitosis. As we have already seen, none of these changes take place if the insect is decapitated; they are all the result of activation by the moulting hormone.[215]

There are other tissues whose growth is not dependent upon the moulting hormone.   In the fat body cells of the *Rhodnius* larva precisely the same sequence of changes occurs after feeding, and may continue until the cells undergo mitosis.   But the growth of the fat body is not affected by decapitation: it is the result of nutrition alone and is not directly influenced by the action of the moulting hormone.[215]   We must conclude that the moulting hormone is not an essential element for catalysing growth in general.   It is a ' messenger ' which calls forth the appropriate growth responses in cells which otherwise would be expected to remain dormant.

The growth of insect tissues in the absence of the moulting hormone is likewise evident during embryonic development (p. 110).   In the egg of *Locustana pardalina* the burst of mitoses at the end of diapause is initiated before the ventral glands ('prothoracic glands') are formed. These glands come into action later, and show their maximum activity at the time when the embryonic cuticle is being moulted.

The growth of the imaginal discs seems to be dependent on the presence of the moulting hormone.   If the ring gland is excised from the *Calliphora* larva the growth of the imaginal buds is arrested.[22]   The same conclusions have been reached by using the adult fly as a culture medium for imaginal discs.   The discs of *Drosophila* cease growing if they are transplanted to a normal adult; but they will continue to grow if the larval ring gland is implanted at the same time.[13]   On the other hand, in the organ discs of *Drosophila*, growth appears to be independent of moulting; for the growth curve in the eye discs rises steadily during larval life, with no sign of breaks related with moulting.   It may be that throughout the early larval stages in *Drosophila* the growth and moulting hormone is always present in the blood; the imaginal discs can thus grow continuously, but the larval epidermis must go through cycles of cuticle deposition and must therefore of necessity show cyclical growth.

Mode of action of ecdysone

During the pupal stage of insects the course of oxygen consumption follows a U-shaped curve, being high at the beginning and end of the pupal stage and low over the middle period.   This curve is associated with a corresponding curve in the level of oxidative (and dehydrogenase) enzymes. Some authors have been inclined to consider the course of enzyme concentration as being the *cause* of the U-shaped curve of oxygen uptake. But the oxygen consumption of the body at any time is determined by metabolic activity—whether this be muscular or secretory activity on the one hand, or the energy required for endothermic chemical syntheses on the other. The level of enzymes present is commonly adapted secondarily to the metabolic needs.[216, 230]

Williams and his colleagues[216] showed that the cytochrome system in the dormant pupa of *Hyalophora* was very small in amount, and cytochrome *c* appeared to be completely absent so that the insect showed extraordinary tolerance to carbon monoxide and to cyanide. Soon after the renewal of growth cytochrome *c* reappeared and it was suggested that synthesis of cytochrome *c* was a key process initiated by the moulting hormone. More recently it has been found that even in the dormant *Hyalophora* pupa respiration is mediated by cytochrome oxidase and that the apparent resistance of the system to inhibitors is due to a great excess of cytochrome oxidase in relation to the very small amount of cytochrome *c* present. The hypothesis that the synthesis of cytochrome *c* is the essential change that restores growth has therefore been abandoned.[72]

The rising consumption of oxygen which accompanies the moulting process is doubtless due to many causes; but it must be due mainly to endothermic syntheses—partly of food reserves for storage, largely of the materials needed by the growing body. The processes of nucleolar enlargement, mitochondrial increase and RNA synthesis, described in the preceding section as characteristic effects of ' activation ' by the moulting hormone, point to an active renewal of growth with the synthesis of protein. It was therefore suggested that the prime effect of the moulting hormone is the restoration of protein synthesis in those tissues that are concerned with growth.[215]

The renewal of protein synthesis is certainly one of the earliest effects detectable in those cells that are concerned with moulting. But it cannot be regarded as a process for which the moulting hormone is always necessary: that is true only of the epidermal cells and some other cells that are concerned in moulting; the effect of the moulting hormone is to arouse these particular cells from their dormant state and thus bring about the renewal of growth. Results pointing to precisely the same conclusion have been reported in the dormant pupa of *Hyalophora*. Respiration and the incorporation of labelled amino acids into protein are very low in dormant pupae; they are greatly increased not only when development begins but also after wounding.[216] Growth and moulting are not the only activities that demand protein synthesis.

When renewed growth of the wing epidermis of diapausing Saturniid moths is induced by ecdysone, the synthesis of all types of RNA is stimulated.[51]

Just where the moulting hormone is acting is still uncertain. It may be that it acts by influencing permeability relations within the cells thus allowing access of enzymes to their substrates.[215] During recent years the belief has been gaining acceptance that the moulting hormone may be acting, not upon some cytoplasmic mechanism but upon the nuclear genes themselves. The main support for this belief comes from the observation of the giant ' polytene ' chromosomes in the salivary glands of the larvae of *Drosophila* or of *Chironomus*. Specific bands, believed to

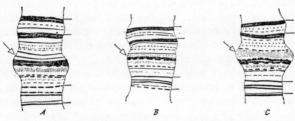

FIG. 7.   The ' puffing ' of the locus 18-C on chromosome 1 in the salivary glands of *Chironomus tentans* after the injection of ecdysone (after Clever) A, untreated control; B, 30 min. after injection; C, 2 hours after injection.

represent particular gene loci in these chromosomes, become enlarged to form so-called ' puffs ' in many cells throughout the body. These changes may appear within fifteen minutes of exposure to ecdysone in some gene loci; within an hour or two at others (Fig. 7).[4, 37] They are interpreted by some authors as evidence that in response to the hormone, specific ' messenger ' ribonucleic acids are being synthesized as a first step in the production of specific enzyme proteins needed for growth and moulting.

Likewise the action of ecdysone in inducing tanning in the puparium of *Calliphora* and other flies is regarded as an effect at the gene level resulting in synthesis of the necessary enzymes (p. 21).

Moulting hormone, mitosis, growth and differentiation

It is sometimes claimed that the induction of mitosis is an important effect of the moulting hormone; but this varies in different tissues and in different insects. In the integument of *Locusta* a wave of mitoses seems to be an integral part of the moulting process: a few hours after the completion of one moult there are nearly 8,000 nuclei per mm²; mitosis begins within a day or two and when it is complete there are more than 12,000 nuclei per mm^2.[216] On the other hand in the abdominal integument of the *Rhodnius* larva mitosis is proportional to the degree of stretching of the body wall and seems to be induced by the mutual separation of activated cells. If moulting is produced experimentally (by the injection of ecdysone) in the unfed larva, the cells are already very crowded and moulting takes place with virtually no mitoses. Clearly, induction of mitosis is not a direct effect of the hormone.[217]

It is also sometimes said that the moulting hormone promotes differentiation (it has been called the ' growth and differentiation ' hormone). But, as we shall see later, differentiation results from the mutual relations of epidermal cells (Chapter 7); the hormone cannot be said to *cause* differentiation. On the other hand differentiation occurs only among cells in process of growth and moulting; and since growth and moulting are due to the moulting hormone, differentiation can be regarded as a consequence of its activity.

If the level of oxygen in the environment is low, or if the temperature falls below a certain level, growth is slowed down or even arrested altogether. But the more interesting effect is that of high temperature. If the temperature to which *Rhodnius* is exposed is raised above 35°C moulting is arrested; although other activities continue unchanged up to a temperature of about 40°C. It has been proved that this arrest can be due to interference with the secretion of the activating hormone from the brain, interference with the secretion of the moulting hormone by the prothoracic glands, and interference with the epidermal cells themselves.[214] Detailed analysis of the effect has shown that in each case the high temperature is causing an arrest of protein synthesis.[151] These observations illustrate once more the central importance of protein synthesis in the physiological effects of the growth hormones.

Specialized actions of ecdysone

Hormones are essentially ' messengers ' which serve to inform a given cell or tissue that the time has come to put into operation one or other of the activities with which that cell or tissue is endowed. This endowment derives first from the universal complement of genes carried in the nucleus, and secondly from the ' differentiation ' that the cell or tissue has undergone, which renders it able, or ' competent ', to manifest a particular genetically determined activity. In differentiation there is therefore a temporal element: the cell may be ready to manifest a characteristic activity only at a particular time in development.

These general principles can be illustrated by two specialized effects of ecdysone. The first concerns the higher Diptera, such as *Musca* or

PLATE II

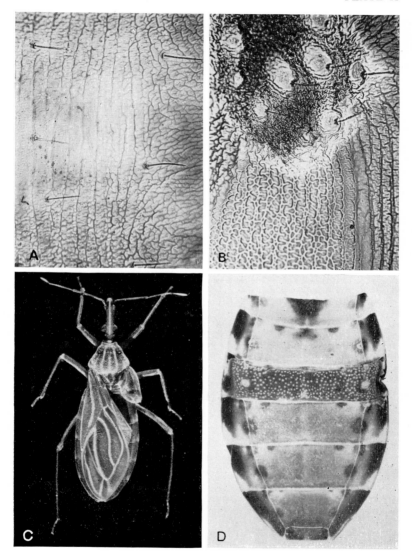

PLATE II. A. Normal abdominal cuticle in adult *Rhodnius*. B. The same with small patch of larval cuticle (with stellate folds and smooth rounded plaques) overlying an implanted corpus allatum. C. Adult *Rhodnius* with a larval wing on the right side following local application of juvenile hormone in the 5th-stage larva. D. Dorsal integument of the abdomen of an adult *Rhodnius* with one larval segment.

PLATE III

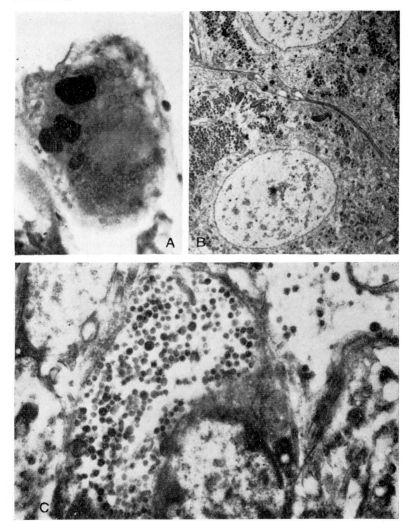

PLATE III. A. Section of pars intercerebralis of *Rhodnius* showing neurosecretory cells stained with chromehaematoxylin. B. Low power electron micrograph of neurosecretory cells in adult female *Calliphora* showing masses of neurosecretory granules. (Courtesy of E. and M. Thomsen.) C. Electron micrograph of corpus cardiacum of *Carausius* showing swollen ending of axon with neurosecretory granules. (Courtesy of G. F. Meyer.)

Calliphora, in which the delicate pupa remains within the larval skin. The larval cuticle is not cast off, but shortly before the pupal transformation occurs it is ' tanned ', hardened and darkened to form a protective ' puparium ' (Fig. 8). The biochemistry of this change has been worked out in detail in *Calliphora* by Karlson and his colleagues[92] (Fig. 9). Tyrosine, present in quantity in the circulating haemolymph, is taken up into the epidermal cells, oxidized to ' dopa ' (dihydroxyphenylalanine) which, under the action of ' dopa '-decarboxylase is converted to dopamine and then acetylated to form acetyl-dopamine. This is oxidized to the corresponding quinone, which acts upon the protein in the cuticle and tans it, causing it to polymerize into the hard, dark, hydrophobic material ' sclerotin '. The key enzyme that is needed to

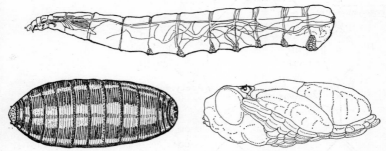

FIG. 8. The larva of *Musca* above. To the left, the hardened larval skin (puparium) enclosing the pupa. To the right, the pupa exposed after peeling away the puparial shell (after Thomsen).

initiate this process is ' dopa '-decarboxylase and this is produced in the epidermal cells under the action of ecdysone.

The synthesis of this enzyme protein is effected by a specific ' messenger-RNA ' attached to the ribosomes in the cytoplasm. There is evidence that this messenger-RNA is the product of a specific region of the chromosomes in the nucleus, where a specific ' RNA-polymerase ' is formed at a fixed site on the chromosomal DNA. This site is one of the chromosomal regions which are activated and undergo ' puffing ' under the action of ecdysone (p. 19). Thus, according to the theory developed particularly by Karlson and his colleagues, ecdysone exerts a specific action directly upon this gene locus, induces the formation of a specific RNA-polymerase and hence the specific messenger-RNA which passes into the cytoplasm and leads to the synthesis of the key enzyme dopa-decarboxylase and thus induces puparium formation.[96] This response can be obtained even when isolated nuclei from epidermal cells are incubated *in vitro* with ecdysone.[48]

The alternative interpretation is that ecdysone has a more general effect on the cells; that it acts as a messenger or stimulus which causes the cell to set in motion those biochemical changes that are appropriate to the state of differentiation at that time and place; but that the precise nature of this stimulus is still unknown. (An effect on permeability properties within the cell or its nucleus is one hypothesis.)[215] This way of thinking lacks the attractive precision of the theory of specific action

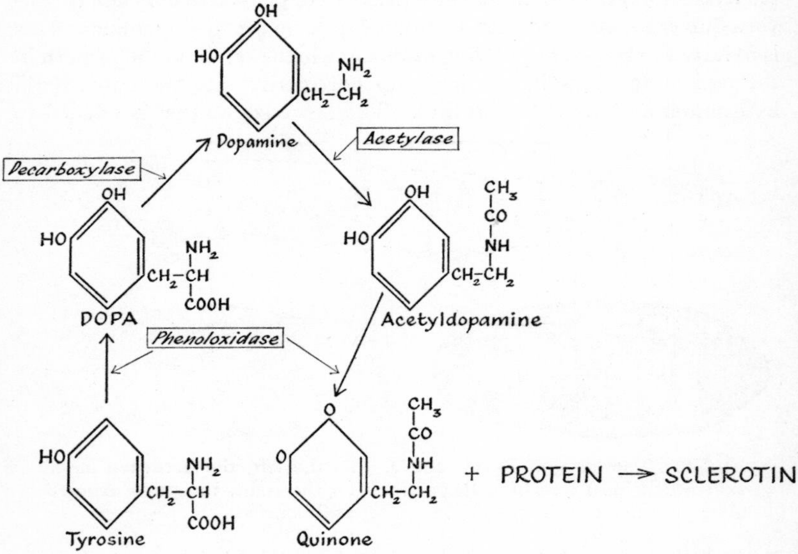

FIG. 9.   The biochemistry of sclerotin formation in the larva of *Calliphora* (after Karlson and Sekeris).

on gene loci as outlined above, but it will account better for some of the difficulties attaching to this theory. Thus, although the time relations between exposure to ecdysone and the formation of dopa-decarboxylase are quite clear-cut in the full-grown maggot of *Calliphora*, there is *no* rise in dopa-decarboxylase during the peak of production of ecdysone in the pupa of *Calliphora* when it is developing into the adult. On the other hand, dopa-decarboxylase *is* produced in adult flies shortly before emergence (in readiness for cuticle hardening) although, at that time, ecdysone cannot be detected. It is therefore possible to regard ecdysone just as a 'signal substance' which indicates to all parts of the body that growth and moulting are to proceed. Precisely what elements in the gene system are activated, and what processes are initiated in particular cells, will differ in different tissues and at different stages of growth.

The second example of a specialized effect of ecdysone that has been closely studied is the colour change that takes place in certain caterpillars before they settle down to pupate. A well-known example is the puss moth caterpillar *Cerura vinula*. While it is feeding on the foliage of willow or poplar, this caterpillar is bright green with a saddle-shaped brown pattern on the back. Twelve days before pupation the larvae become chocolate brown in colour from the deposition of ommochromes in the epidermal cells, and they discharge a red faecal pellet. Such larvae are far less conspicuous when they wander down the trunk and over the earth than they would be if they retained their green coloration. This change is brought about by ecdysone: a very small dose of ecdysone will cause reddening of the epidermis and the change in behaviour; a larger dose reddening of the fat body; whereas a very large dose will induce the pupal moult in a ligated caterpillar directly without any colour change.[21]

In this instance no attempt has been made to ascribe these colour changes to the direct intervention of ecdysone at the level of the gene. In biochemical terms the red coloration derives from the fact that the stored proteins are being mobilized and hydrolysed for use. They provide an excess of the rather toxic amino acid tryptophan, and in insects this is eliminated by conversion into ommatin pigments which may be stored within the cells or discharged in the faeces.[112]

Chemistry of the growth hormones: the brain hormone

The nature of the product of the neurosecretory cells, the features of which can be studied with the electron microscope in the cell bodies and as it passes down the axons, will be considered in a later chapter (p. 88). Here we are concerned with the 'activation hormone' which circulates in the blood and stimulates the prothoracic glands to liberate the moulting hormone. The chemical nature of this material is still uncertain.

Most experiments have been made on the pupa of the silkworm *Bombyx* and on other Lepidoptera. The brain is removed from a young pupa and development is arrested. The brainless pupa becomes a permanent or 'dauer'-pupa. It retains the prothoracic glands but these are not activated. On introduction of the 'activation hormone' the prothoracic glands are induced to secrete the moulting hormone and development is renewed. In this system, however, the activity of the prothoracic gland seems to be only temporarily suppressed. Eventually, after some months, the glands become active spontaneously. It therefore seems possible that, in experiment, various non-specific agents might serve to activate the gland and thus be mistaken for the activation hormone. Implantations of corpora allata have this effect; and so do oily extracts with juvenile hormone activity (p. 68). Likewise, an oily extract from the pupal brain of *Bombyx* will cause the brainless ' dauer '-pupa to resume development; and even purified cholesterol isolated from this source, or from other sources, will have the same effect.[101] It is, however, difficult to believe that cholesterol itself can be the active substance, for the amount of cholesterol normally present in the haemolymph of the diapausing cecropia pupa is about 10,000 times that needed to stimulate the prothoracic gland.[65]

In most vertebrate animals the active product of the neurosecretory cells seems generally to consist of proteins or small polypeptides. Indeed Ichikawa and Ishizaki have prepared aqueous extracts from the brain of *B. mori* which are highly active in inducing adult development in the test pupae; and the active material seems to be a protein. It can be extracted with saline, and is precipitated by ammonium sulphate and trichloroacetic acid; in solution it is non-dialysable and is inactivated by bacterial proteinases. The active principle has been purified to a considerable extent and seems to be a mixture of heterogeneous peptides with molecular weights ranging from 9,000–31,000.[85]

Chemistry of ecdysone

It was soon observed that the moulting hormone of insects is non-specific: the blood of moulting *Rhodnius* will induce moulting in *Cimex* and *Triatoma*[211] and blood from developing *Hyalophora* will cause renewed growth in dormant pupae of *Telea*; extracts from *Calliphora* will cause puparium formation in *Drosophila*[212] and extracts from pupae of *Galleria* are effective in *Calliphora*.[212]

The non-pupating posterior halves of *Calliphora* larvae that had been ligatured shortly before puparium formation have served as a useful test material for the moulting hormone. A very small quantity of the hormone injected into these fragments will induce hardening and darkening of the cuticle. Methods for concentrating and isolating the active factor were worked out by Butenandt and Karlson.[24] The same methods were applied to developing pupae of the silkworm *Bombyx* which could be obtained in large quantities from the silk industry. Starting with 500 kg of silkworm pupae Butenandt and Karlson succeeded in isolating 25 mg of the active substance in the form of needle-like crystals. It is this material that was named ecdysone (α-ecdysone). It was shown to be associated with smaller quantities of a more polar and slightly more active substance provisionally named β-ecdysone.

The chemical nature of these materials took about ten years to elucidate. They were soon shown to be nitrogen-free and ecdysone was eventually found to have a molecular weight of 464 and the empirical formula $C_{27}H_{44}O_6$. The application of novel X-ray crystallography methods, devised by W. Hoppe, directly to crystals of pure ecdysone, showed that the fundamental structure was that of a steroid. The full structure of ecdysone was finally elucidated by Huber and Hoppe, and it has since been synthesized.[92] β-ecdysone proved to be the same compound with an additional hydroxyl group: 20-hydroxyecdysone. This

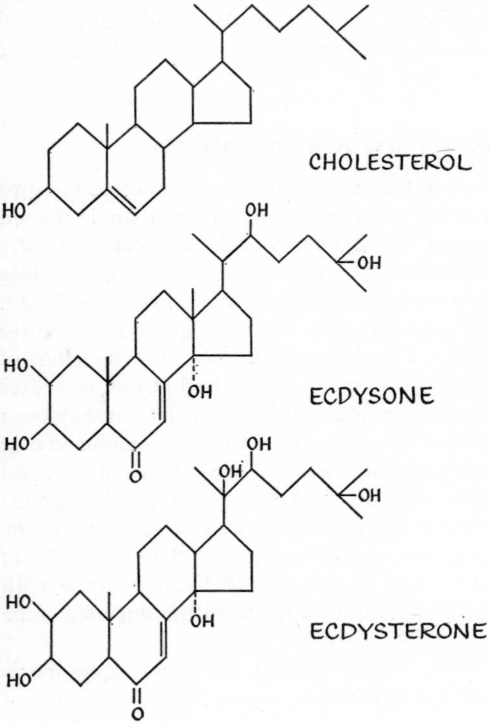

CHOLESTEROL

ECDYSONE

ECDYSTERONE

FIG. 10.   The structure of ecdysone in comparison with cholesterol.

appears to be identical with the factor isolated from crustacea ('crust-ecdysone') and the material named 'ecdysterone'. The structure of these compounds in comparison with cholesterol is shown in Fig. 10.

One reason why it took so long before the steroid nature of this hormone was recognized was because the large number of hydroxyl groups on the molecule rendered the material readily soluble in water. During recent years it has been found that steroids of this type are abundant in certain plants, where they appear to have been overlooked for the same reason. Thus the foliage of yew (*Taxus*) is a rich source of ecdysterone: 1 g of dry leaves yields 500 μg of pure ' ecdysterone '—indeed 70 g of the dried leaves will yield as much active substance as 1,000 kg of *Bombyx* pupae![186] The rhizomes of bracken (*Pteridium*) are another source of several ecdysone-like compounds; and there is a long list of plants, notably Japanese plants, which have yielded a wide variety of new ' ecdysones '.

The relation between ecdysone and cholesterol naturally raised the question whether ecdysone could be derived from cholesterol in the diet. It appears that this may well be so, for when Karlson and Hofmeister[94] administered tritiated cholesterol to *Calliphora* larvae they were later able to extract labelled ecdysone.

Quantitative requirement for ecdysone

In the ' *Calliphora* test ' it was found that no more than 0·0075 μg of crystalline ecdysone was needed to induce pupation in a single ligated larval abdomen.[216] But these larvae had doubtless secreted already almost enough hormone to induce puparium formation; the test dose was needed only to complete the process. If the injection was delayed, activation of the tissues was gradually lost, and then much larger doses of the hormone were needed. Likewise in *Rhodnius*, the injection of ecdysone has usually been made in larvae decapitated at one day after feeding; but by that time a certain amount of hormone has already been secreted. Decapitated fourth-stage larvae of *Rhodnius* (weight 80 mg) receiving 0·25 μg of ecdysone show activation of the epidermal cells and some development of muscle fibres in the ventral abdominal muscles; 0·5 μg causes more development, but growth comes to an end before the new cuticle is laid down; 0·75–1·0 μg will lead to complete formation of the new cuticle. This is a dose of about 10 μg/g[214] which agrees with the dose needed to induce complete development in the dormant pupa of *Hyalophora*.[24]

In the caterpillar of *Cerura vinula* (p. 23) 0·5 μg of ecdysone is needed to cause reddening of the epidermis; 2·5 μg causes reddening of the fat body; whereas 22–44 μg are necessary to induce the pupal moult directly in the ligated larva.[21] This last value is of the same order as that needed in *Rhodnius* and *Hyalophora*.

Estimates of the extractable ecdysone in the full-grown larva of the fleshfly *Sarcophaga* have shown that although the amount necessary to induce puparium formation is very small (about 0·035 μg of α-ecdysone) at no time is there sufficient circulating in the blood to have this effect. In other words, under natural conditions, the tissues are being furnished with a continuous small supply.[150] In *Calliphora* any excess of ecdysone introduced into the blood is rapidly inactivated by enzymes in the fat body: within 2 hours 80 per cent can be lost.[93]   It is generally assumed that ecdysone is the product of the prothoracic gland; but one slightly puzzling fact is the relatively large amount of ecdysone that can be extracted from adult insects such as *Bombyx*[179] and the grasshopper *Dociostaurus*.

2: Hormones and Arrested Development

It is characteristic of many insects that at certain seasons of the year, usually during the winter, sometimes during the hot dry season, they are subject to a period of developmental arrest, or ' diapause '. This state is associated with a low rate of metabolism. In overwintering pupae, for example, the rate of oxygen consumption at the time of pupation is quite high; it soon falls and throughout the winter is very low, but rises again as the time for emergence of the adult approaches in the spring. In the diapausing pupa of *Hyalophora cecropia* the oxygen consumption is only 1·4 per cent of that in the mature larva, and 5 per cent of that in the adult just before emergence. By the first day of visible adult development the oxygen consumption is increased to $3\frac{1}{2}$ times that of the diapausing pupa; it then rises gradually to 20 times that value.[216] This U-shaped curve of oxygen uptake retains the same form even at a constant favourable temperature; it is the same curve in exaggerated form as is seen in uninterrupted pupal development (p. 17).

In the past, attention was focussed on the low rate of metabolism during diapause, and the diapause state was commonly described as being the *result* of the depressed metabolism. The depression was attributed to a hypothetical ' diapause substance ' or ' Latenzstoff ' which actively inhibited metabolism; and Roubaud, in particular, developed a theory according to which accumulating excretory products suppressed metabolism and led to a state of developmental fatigue, or ' asthenobiosis ', which was the immediate cause of diapause. Only after a period of metabolic rest at a low temperature, during which the toxic excretory products were eliminated, could metabolism increase once more and development be resumed.

We saw in the preceding chapter that *Rhodnius* larvae which take repeated small meals of blood fail to *grow* because they lack the necessary stimulus of abdominal distension; and insects decapitated soon after feeding fail to grow, even though they have received a full-sized meal,

28

because they lack the moulting hormone.   " Such insects are in a state of diapause. Growth and moulting can take place in them only if the requisite hormone is introduced into the blood. And this naturally suggests that diapause in other insects (when it is not brought about by the *direct* effect of the environment) may result from the temporary failure of growth-promoting hormones, due sometimes perhaps to an inborn rhythm, sometimes perhaps to the indirect effect of environmental factors."[211] These observations provided the basis for an alternative theory of diapause.

Hormones and natural diapause

This theory that natural diapause may be due to the lack of the hormones necessary for growth was first investigated experimentally in the pupa of the giant silk moth *Hyalophora cecropia*. *Hyalophora* enters diapause immediately after pupation and, as in many other insects, diapause is brought to an end by exposure at 3–5°C for several months. Williams joined ' diapause pupae ' in parabiosis with ' chilled pupae ' whose diapause had been terminated by prolonged cooling. The development of the chilled pupae was not arrested; there was no evidence of the presence of an inhibitor of development in the diapausing pupa. On the contrary, the diapause pupae were induced to develop as soon as development in the chilled pupae began. As in *Rhodnius* (p. 3) the interchange of haemolymph through a connecting glass tube was sufficient to initiate growth.[212] Implants of various organs removed from chilled pupae confirmed that only the brain would terminate diapause; it is the brain that is reactivated by cold.[212]

The diapausing pupa of *Hyalophora*, as already described (p. 4), has proved most valuable material for studying the action of the growth hormones of insects. Like the larva of *Rhodnius* it will not resume its growth unless the moulting hormone is present. Soon after the ending of diapause, the prothoracic gland is activated (as we have seen) and then a pupa in diapause can be induced to develop if such an activated prothoracic gland is implanted into it. The prothoracic glands are necessary for development in the pupa for six days after the termination of diapause. Thereafter, enough hormone has been secreted for the moulting process to continue in the absence of the glands.

In view of the relations between these two sources of hormones, the brain and the prothoracic glands, it is not surprising that removal of the brain has a different effect on diapause in different species of insects. Certain species, such as the summer broods of *Arctias luna* and *Arctias selene*, do not have a pupal diapause. But if the brain is removed immediately after pupation, no further development occurs. That is because, although the brain remains active, the prothoracic gland is temporarily inactive.[111]   In the alder fly *Sialis* the prothoracic gland in the full-grown larva appears to have been already activated by the brain in the autumn.   The larva remains in diapause because, at the low winter temperatures, the moulting hormone is simply stored.   It is liberated on warming, so that in this insect pupation will occur, even in the decapitated larva, when it is brought into warm surroundings.[216]

Another insect in which the hormonal changes during diapause have been closely studied, notably by Church, is the wheat-stem sawfly *Cephus cinctus*.[216]   The sawfly larva over-winters in diapause in a small chamber in the wheat stub.   In this larva visible imaginal development continues very slowly even during diapause.   As in *Drosophila*, *Hyalophora*, *Rhodnius*, etc., the brain hormone is needed to sustain the activity of the prothoracic glands even after these have begun to secrete. If the secretion is insufficient, a high temperature ($35°C$) will cause the post-diapause larva to revert to diapause.   The heat treatment halts thoracic gland activity (p. 20) and the gland reverts to the inactive state.[216]

These experiments provide strong support for the hormone theory of diapause.   They do not exclude the possibility that a ' diapause factor ' may exist which inhibits the activity of the endocrine glands responsible for growth.   Such a factor has often been sought without success (but see ' Diapause in the silkworm egg ', p. 38).   From time to time it has been suggested that a secretion from the corpus allatum may be responsible for keeping the insect in diapause.   The most detailed support for this suggestion comes from experiments with the diapausing larva of the rice-stem borer *Chilo*[216]—but these experiments are susceptible to more than one interpretation.[216]

It must, however, be realized that there are many limiting factors for growth in insects.   Mosquito larvae, *Aëdes*, which are not permitted to fill their tracheal system cease growing in the second instar from lack of *oxygen*; and *sodium chloride* may likewise be a limiting factor in the growth of mosquito larvae.   *Rhodnius* larvae deprived of their symbiotic actinomyces cease growing in the fourth or fifth instar from lack of *vitamins*; diapause in larvae of the pink bollworm *Platyedra gossypiella* may be induced by lack of water in the food and may persist two years or more; but pupation occurs within a few days after moistening.[111]   In the wheat-stem sawfly *Cephus*, after the termination of larval diapause in the spring, it may be reinduced by partial desiccation.   It is evident that many different factors may be concerned during natural diapause, both in the direct arrest of growth and in suppressing the secretion of growth hormones.

Induction of diapause

Growth in many insects is influenced by what is termed the ' group effect '. There is an optimal density for the development of *Tenebrio* larvae in culture: a moderate degree of mutual stimulation is beneficial; larvae in complete isolation show delayed growth; but above a certain density development is retarded.[212] In *Ptinus tectus* the adverse effect of stimulation is the more evident. If a group of several larvae is reared in one container development may be prolonged by as much as 40 per cent. For example, with a group of eight larvae as much as 5 g of food per larva is required before the effect disappears; that is 100 times as much as is needed for full and speedy development of an isolated larva.[212] The maximum rate of growth in *Blattella* is seen when there is a space of 14 cc per insect.[212] Perhaps these are all examples of the influence of nervous stimulation on hormone production and serve to illustrate the central importance of the brain in the control of growth.

In many insects diapause at some fixed stage in the life cycle is obligatory; it supervenes irrespective of the environmental conditions. That is so in the case of pupal diapause in many Lepidoptera from temperate latitudes. Diapause is built into the inherited pattern of development. Other insects, such as the blowfly *Lucilia* will breed continuously if conditions are uniformly favourable; diapause in the larva is brought on by such adverse conditions as poor food, drought, cold or excessive moisture. But development does not proceed automatically upon return of the larva to a good environment; the arrest may be prolonged for weeks.

The intensity of such an arrest varies widely in different insects; and in different localities strains have been evolved which differ in their propensity for entering diapause. Some strains in the European corn borer *Ostrinia nubilalis*, for example, are univoltine: they have a single generation in the year and the full-grown larva regularly goes into a diapause that persists all winter. Another strain is bivoltine: a second summer generation develops without a diapause and only the autumn generation gives rise to dormant larvae.[218] Likewise the European spruce sawfly *Gilpinia polytoma* overwinters in the cocoon; and some of the resting larvae may lie over for seven years. After a rest at a low temperature, contact with water provides the stimulus for development. Here again there are univoltine and bivoltine races with different geographical distribution.[218]

In many insects, in which diapause appears to be an integral part of the developmental pattern, uninfluenced by the environment, it has been found that the arrest is indeed induced by environmental stimuli but that these are experienced so long before the arrest becomes apparent that in the past they have been overlooked.   A moment's reflection shows that this is a desirable arrangement: an insect will wish to know many weeks in advance that winter is coming, in order that its own development or the development of its offspring may be arrested during the adverse season when frost may imperil a feeding insect or food may be absent altogether.   In this type of facultative diapause the most important advance information is derived from three sources: nutrition, temperature and day-length.

Euproctis phaeorrhoea pass the winter as young larvae.   These larvae enter diapause if they are fed on ageing foliage; they fail to do so if fed exclusively on young foliage.   In *Loxostege sticticalis* diapause occurs in the full-grown larva; it can be induced by exposing the young larva, at quite an early stage of growth, to a restricted period of unfavourable nutrition.   In the rice green caterpillar *Naranga aenescens* diapause supervenes in crowded cultures; but this appears to result from mutual stimulation and not from deficient nutrition.   In the pupa of the corn earworm *Heliothis armigera* diapause is caused by low temperature during the larval period and in many other insects the incidence of diapause brought on by unfavourable nutrition or day-length may be countered by a rise in temperature.

But the most important factor in the induction of diapause is the length of day.[221]   Many species of Lepidoptera require a certain critical length of day during larval life if diapause is to be prevented and un-interrupted pupal development is to occur.   Some examples are shown in Fig. 11.   In *Acronycta rumicis* or in *Diataraxia oleracea* the light period must be longer than 16 hours; in *Pieris brassicae* and *Grapholitha molesta* 13–14 hours is sufficient.   As the day-length is shortened to perhaps 6 hours or less, diapause is again partially eliminated.   In some insects, such as *G. molesta*, all the pupae develop without diapause if the larvae have been kept in total darkness; in *A. rumicis* about 80 per cent develop after this treatment, in *D. oleracea* about 20 per cent.   In *Ostrinia nubilalis* diapause is induced in the larva over a narrow range of photoperiods of 10–14 hours of light; and a corresponding period of 12 hours of darkness is equally necessary.   Indeed, in many insects the length of the dark period is more important than the duration of the light: photoperiod is really a misnomer.[111]

In the photoperiodic chart (Fig. 11) the left-hand half of the figure, which illustrates reactions to day-lengths below 10 hours, has, of course, no ecological significance, since such photoperiods occur in winter only in high latitudes and do not concern insects at the feeding stage.   The results illustrated are non-adaptational; hence the wide and unpredictable variations in different insects.   It is the right-hand end that relates to ecological conditions; in the region of the critical threshold changes of less than one hour in day-length lead to complete inversion of the type of development.   The precision of this response is the result of rigorous selection.[42]   Most insects studied are sensitive in the photoperiodic reaction to only the short-wave part of the spectrum (blue-green to near ultraviolet).   This is just the opposite to plants and vertebrates; it suggests that carotinoids, which have an absorption spectrum in this region, may be concerned in the reaction.[42]

In *Pieris brassicae* the long photoperiod (e.g. 16 hours of light)

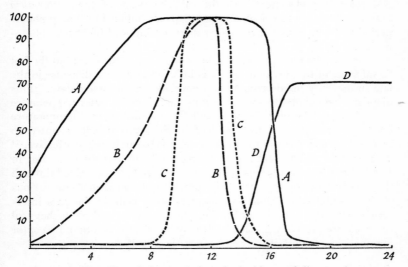

FIG. 11.   The effect of photoperiod on the incidence of diapause in some species of Lepidoptera.   Ordinate: percentage of individuals entering diapause.   Abscissa: hours of light in 24 hours.   A, *Acronycta rumicis* (after Danilyevsky).   B, *Grapholitha molesta* (after Dickson).   C, *Pyrausta nubilalis* (after Beck).   D, *Bombyx mori* bivoltine race (after Kogure).

acts directly on the brain itself to maintain secretion of the brain hormone and so to prevent the onset of diapause; whereas a short photoperiod (e.g. 8 hours of light) promotes the onset subsequently of the pupal diapause.   The larva first becomes sensitive after the third moult.   Until that time the head capsule is entirely black; at the third moult the clypeus forms a yellow triangle which admits the passage of light to the brain.   It is possible to take a brain from a larva during the light-sensitive period and implant it into the abdomen of a larva which has passed the light-sensitive stage and, having been exposed to short days, would normally undergo a pupal diapause.   The implanted brain can now be controlled by the conditions of illumination: if the host larva is exposed to short-day conditions it goes into a pupal diapause; but if exposed to a 16-hour photoperiod diapause is prevented.[34]

In many of these insects, such as *Pieris, Ostrinia, Acronycta,* and doubtless many more, there are geographical races which are adapted to go into diapause at day-lengths appropriate to the latitudes at which they occur. Thus in *Acronycta* the limiting photoperiod ranges from nearly 20 hours at Leningrad in the north to $14\frac{1}{2}$ hours on the Black Sea coast in the south.[42] In the cabbage moth of Japan *Barathra brassicae* there is a transient summer diapause induced by a long photoperiod and a long persisting winter diapause induced by a short photoperiod.

Most insects respond to a photoperiod of defined length; but the last larval stage of the dragonfly *Anax* is induced to go into diapause in the late summer by the decreasing photoperiods on successive days. This diapause soon comes to an end; but by that time the water temperature has fallen and the larva is then held ' quiescent' by low temperature until the spring. The threshold of sensitivity is usually adapted so that the light of the full moon (0·01–0·05 foot-candles) is ineffective.[111] The light is usually said not to be received through the eyes, but in *Dendrolimus pini* the photoperiodic stimuli seem to be perceived through the ocelli of the caterpillar. The precise link between these various environmental stimuli and the neurosecretory cells in the pars intercerebralis of the brain is not known. Both nervous and photochemical reactions are probably involved. It may be that the carotinoid receptor pigments reside within the neurosecretory cells themselves.

Termination of diapause

When dormancy is the result of the direct depression of metabolism by low temperature or by partial desiccation, or is the consequence of the suppression of protein synthesis by high temperature (p. 20), it is commonly brought to an end as soon as the normal environmental conditions are restored. Here we are concerned with the state of so-called ' true diapause' in which dormancy persists even in a favourable environment. In the familiar winter diapause of insects from temperate climates it has long been known that the dormant insect is ' reactivated' by a more or less prolonged exposure to cold, and the nature of this curious response has been the subject in the past of much speculation.

Diapause is to be thought of as being essentially a physiological mechanism for survival during an adverse season, most commonly the winter cold. The mechanism seems to consist in the separation of physiological processes, some of which require a low temperature for their completion, while others require a high temperature. It seems that certain processes, called ' latent development ', or ' diapause development ',[216] have become adapted by natural selection to proceed only in some low temperature range characteristic of the species; while the main process of development will go forward only in the upper range of temperature.

Each of these two processes has a temperature curve of the usual type, but with widely separated optima.   In Fig. 12 this is illustrated in the egg of the Australian grasshopper *Austroicetes*.   When the temperature curves for the two processes do not overlap (as happens in the egg

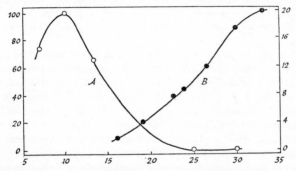

FIG. 12.   The influence of temperature on diapause development (curve A) and post-diapause development (curve B) in the eggs of *Austroicetes cruciata* (after Andrewartha). Ordinate: to the left, percentage of eggs which complete diapause development during 60 days at the temperature specified; to the right, relative rates of development of post-diapause embryos at different temperatures. Abscissa: temperature in °C.

of the silkworm) *no* development occurs at an intermediate temperature; and when the two curves come close together the existence of diapause may be difficult to detect.   This kind of variation is well seen among the Saturniid moths *Philosamia ricini*, *P. cynthia*, *Antheraea pernyi* and *Saturnia pavonia*, where the degree of overlap varies in accordance with the range of temperature to which they are exposed in normal life.[42, 111]

The optimum temperature for ' diapause development ' is indeed related with the temperature normally experienced by the species:[42] say 10 to 20°C for subtropical species, 0 to − 10°C in temperate regions. Thus diapause development can occur at − 5°C in eggs of *Malacosoma disstria*, at − 10°C in larvae of *Gilpinia polytoma*, − 15°C in pupae of

Saturnia pavonia. In some sawflies the eonymph needs at least 4 months at 0°C and will lie over for a second winter if the duration of cooling is insufficient. Pupae of the moth *Biston* have been known to pass through 7 years before resuming their development. In the egg of the wheat-bulb fly *Leptohylemyia coarctata* it seems that different temperatures are required for different stages of the diapause process— but a temperature of −24°C is highly effective.

That is the usual state of affairs; but when diapause occurs in the hot dry season, ' diapause ', or ' latent ', development may require a high temperature for its completion. That is so in the moth *Diparopsis castanea* from tropical Africa where the optimum temperature for diapause development is 28°C. In *Barathra brassicae* in Japan the summer dormancy is completed at higher temperatures than that in winter.

There is no reason to suppose that the same physiological processes are always involved in ' diapause development '. In post-embryonic diapause it seems usually to be some process in the brain which restores competence to the neurosecretory cells. In hibernating larvae of *Mormoniella (Nasonia)* it looks as though at low temperatures (5°C) the brain hormone accumulates, at high temperatures (25°C) it may be broken down again.[216] This same inactivation of hormone at high temperature (35°C) is seen in *Cephus* larvae coming out of diapause (p. 14) and in *Rhodnius* exposed to temperatures (35°C or higher) above the normal physiological range (p. 20).

In the pupa of *Hyalophora* the brain in diapause seems to be electrically inexcitable; cholinesterase is indetectable and acetylcholine accumulates. When electrical activity and cholinesterase reappear the neurosecretory cells release their hormone and diapause comes to an end. But these changes are associated with growth and development of the nervous system and not with the mechanism of induction and termination of diapause: they occur equally in non-diapausing species.[184] It has been suggested by several authors that in some insects a secretion from the corpus allatum keeps the neurosecretory cells inactive; but the evidence for this view is inconclusive.[216] In *Ostrinia* specialized cells of the hind-gut appear to set free into the blood a hormone (' proctodone ') which is necessary to activate the neurosecretory cells of the brain and thus bring diapause to an end.[3]

Ostrinia and *Pectinophora* are insects in which no chilling is needed; diapause is ended by reversal of the photoperiod. The same is true of *Antheraea pernyi* the pupae of which overwinter in the cocoon in diapause. In studying the photoperiodic response of these pupae Shakhbazov (1961) called attention to the transparent facial cuticle which overlies the pupal brain. When this zone was coated with black paint the pupae behaved as if they were in continuous darkness. Shakhbazov concluded that the light is transmitted through both the cocoon and this facial window to act on some organ in the pupal head. This insect has been studied in detail by Williams, Adkisson and others.[225] They showed that short-day conditions (12-hour photoperiod) inhibit the termination of diapause; long-day conditions (17-hour photoperiod) promote termination. The light stimulus acts directly on the brain: the endocrine system in the neurosecretory cells seems to be responsible. The effective wave length extends for 398–509 $m\mu$, the same spectral region as that of carotinoids, and an intensity of less than one foot-candle is sufficient to activate the system fully.

Diapause in the eggs of grasshoppers

This book is concerned with insect hormones. We have seen that diapause in the post-embryonic stages of insects is associated with and is presumably caused by the lack of growth-promoting hormones. But the growth hormones merely have a controlling or coordinating influence; or they may act as time signals for the initiation of growth; they are not a fundamental requirement for growth as such. There is no evidence that growth hormones operate during early embryonic growth (p. 110); yet in other respects there is no essential difference between embryonic and post-embryonic development. This similarity extends to the physiology of diapause.

Insect eggs at an early stage of development may become dormant as the result of desiccation, as happens in the eggs of Aleurodidae that have been implanted on the leaves of wilting plants; or in the eggs of *Sminthurus* (Collembola) or the South African locust *Locustana pardalina* severely desiccated soon after laying. *Locustana* eggs have survived in dormancy $3\frac{1}{2}$ years; they quickly develop and hatch if they are moistened. But other insect eggs undergo a regular winter diapause resembling that described in pupae. Two examples will be discussed in order to illustrate these similarities: diapause in the eggs of Orthoptera and in the eggs of the silkworm *Bombyx mori*.

Among the Orthoptera the most detailed studies have been made on the eggs of the North American grasshopper *Melanoplus differentialis*. Following upon the observations of Runnström on the changes in the activity of the cytochrome system of the sea-urchin egg at fertilization, Bodine and his colleagues[212] observed that oxygen uptake in the egg of *Melanoplus* was almost completely insensitive to carbon monoxide and to cyanide during diapause; but that the more active respiration before and after diapause was highly sensitive—whence they concluded that mitosis, growth and differentiation are associated with a functional cytochrome system. They failed to find any close correlation between the intensity of respiration and the amount of cytochrome oxidase;[212] it appeared that some other component of the cytochrome system was the variable factor.

The situation seems to parallel closely that described by Williams and his colleagues in the diapausing pupa of *Hyalophora* which has been discussed already in connection with the action of the moulting hormone (p. 18). As we have seen, the apparent resistance of the respiratory system during diapause to such inhibitors as cyanide and carbon monoxide is due to the great excess of cytochrome oxidase in relation to the small amount of cytochrome *c* present.[72] There seems to be no qualitative change in the nature of the respiratory enzymes.

We have here an example of apparently identical changes in metabolism associated with the switch from dormancy to active growth: one at an early stage of embryonic development, with no evidence of hormones being involved; the other in an over-wintering pupa in which the process is regulated by the endocrine system. The diapausing egg of *Melanoplus*, like that of *Austroicetes* (Fig. 12) is 'reactivated by cold'. But the change that is taking place during 'diapause development' must be something different from the restoration of competence in the endocrine organs, for neither nervous system nor endocrine organs have been differentiated in these diapausing eggs; and such reactivation may take place even in fragments of the early embryos.[218] There are other insect eggs, such as those of the beetle *Timarcha tenebricosa* and mosquitoes of the genus *Aëdes* in which dormancy with similar characteristics occurs in eggs with the young larva fully developed and ready to hatch. Here the nervous system is certainly involved; it is not unlikely that hormone secretion may also be involved.

Diapause in the silkworm egg; maternal influence on diapause

The classic example of diapause in the egg is afforded by *Bombyx mori*. Silkworm eggs laid in the autumn will not develop immediately even if kept warm; growth is completely arrested at an early stage. As was first shown by Duclaux in 1869, they will not hatch even in the spring if they have been kept warm (15–20°C) throughout the winter; they will complete their development only if they have been exposed to a temperature around 0°C for several months.

Some races of *B. mori* are single brooded or 'univoltine' so that every generation shows a prolonged period of arrest during embryonic development; other races are 'bivoltine' or even 'tetravoltine'; in these there are one or more uninterrupted generations during the summer before the winter generation of diapause eggs is produced.

The diapause eggs are characterized by having the serosa laden with a brown ommochrome pigment.

Voltinism is hereditary; but when the races are crossed, clear-cut segregation does not occur; for the voltinism of the offspring is influenced by the physiological state of the mother. This ' physiological state ' is determined by the environmental stimuli experienced by the mother when she was in the egg stage, or as a young larva. Thus eggs incubated at a high temperature (25°C) tend to produce moths laying hibernating eggs, whereas eggs incubated at a low temperature (15°C) tend to produce non-hibernating eggs. But even more important than temperature is the effect of day-length during the incubation of the eggs or in the very young larva. If the eggs or young larvae are exposed to a short day of 13 hours of daylight or less, the resulting moths lay non-hibernating unpigmented eggs; exposed to a long day of 14 hours or more, they lay dark hibernating eggs (Fig. 11). It will be noted that the effect of day-length is the reverse of that exerted on most Lepidoptera larvae. The ecological reason for this is the very long time interval between the sensitive stage (the egg or young larva) and the stage which will exhibit or fail to exhibit diapause; that is, the egg laid by the resulting adult female. The short days of spring indicate that there will be time for a summer generation of larvae; the long days of summer indicate that by the time the female comes to lay her eggs winter will be at hand.

Whether the eggs go into diapause or not is determined by some influence from the somatic cells of the mother. A batch of eggs from a single female is generally uniform as regards diapause; and if the ovaries of one race are transplated into another race during the larval stage, the eggs from these ovaries always show the voltinism of their new host. The immediate cause of diapause in the egg is in fact the exposure of the ovaries of the mother to a ' diapause hormone ', or ' hibernation substance '. This hormone is produced by neurosecretory cells in the suboesophageal ganglion of the female. But the secretion of this factor is under the control of the brain, acting by way of the oesophageal connectives.[228] The actual source of the diapause factor is probably a pair of conspicuous neurosecretory cells lying just behind the centre of the suboesophageal ganglion. In pupae destined to give rise to non-diapause eggs these cells (diapause factor cells) are invariably laden with a large amount of neurosecretory material staining with azocarmine: this material seems to be stored and not released. By contrast, the diapause factor cells of pupae destined to give rise to diapause eggs are hardly visible owing to the lack of secretory material in the cytoplasm; the secretion is probably released as soon as it is synthesized. In those pupae which produce non-diapause eggs the release of the neurosecretory material into the blood is completely inhibited by the brain through the oesophageal connectives; whereas in those pupae producing diapause eggs release is actively stimulated by the brain[59] (Fig. 13).

The diapause hormone has been considerably concentrated but its chemical nature is not yet known. One of its effects appears to be to increase the permeability of the ovary to various substances, such as the natural derivatives of tryptophan (kynurenine and 3-hydroxykynurenine) which are converted into ommochrome pigments.

One of the most puzzling features of diapause is the fact that

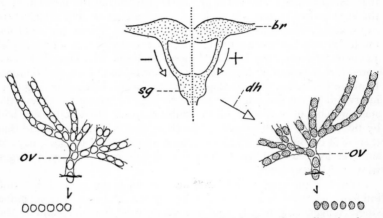

FIG. 13. The role of the brain and suboesophageal ganglion in the control of diapause in the silkworm egg (after Fukuda). To the left: inhibition by the brain leads to production of non-diapause eggs. To the right: stimulation by the brain leads to production of diapause eggs. *br*, brain; *dh*, diapause hormone; *ov*, ovary; *sg*, suboesophageal ganglion.

environmental stimuli may bring about diapause in a developmental stage widely separated in time from the stage on which they act. If females of the parasitic wasp *Nasonia* (*Mormoniella*) are exposed to low temperatures during oogenesis, their progeny will enter diapause at the end of the last larval stage. This effect on the female wears off after a few days and then she begins to produce non-diapausing offspring again. Once the egg has been deposited, no further changes in photoperiod or temperature will modify the occurrence of diapause.[173] A similar effect of low temperature is seen in *Trineptis*; but here the stimulus must act on the larva itself between the second and the final instar if it is to provoke diapause in that generation.[218]

Diapause in the adult insect

Diapause in the adult female insect takes the form of an arrest of egg development. We shall defer consideration of the physiology of this arrest until the hormonal control of oogenesis has been described (p. 84).

3: Hormones and Metamorphosis

In an earlier monograph on the *Physiology of Insect Metamorphosis*[212] an attempt was made to survey the whole subject of metamorphosis in an historical setting. No such attempt will be made in this chapter; but the main outlines of that early history may be briefly summarized as follows. Aristotle had regarded the insect larva as a ' soft egg '. He considered that the embryonic development of insects continued through the ' chrysalis ' stage, right up to the formation of the ' perfect insect ' or ' imago '. William Harvey elaborated this interpretation; he supposed that the insect egg contained so little yolk that the larva was forced to leave it before embryonic development was complete. The larva then proceeded to feed, and to store up reserves of food, until it reverted once more to the egg stage or pupa and resumed its embryonic growth. He considered, therefore, that metamorphosis was a consequence of premature hatching from the egg. This idea was adopted in part by Lubbock (1883) and developed further by Berlese; it is often called today the ' Berlese theory ', according to which the different types of larvae represent the embryonic stage which had been attained by the time of hatching.

Jan Swammerdam (1669) accepted the similarities between pupal development and embryonic growth; but he took issue with Harvey on the grounds that Harvey had written of metamorphosis as a mystical process during which the larval tissues dissolved and were formed anew by some unknown influence ' like wax bearing the impress of a seal '. Swammerdam, however, showed clearly that the structure of the pupa was developed by an orderly process from pre-existing larval structures. He conceived that the adult structure was already latent in the larval body in some obscure form, the nature of which was unknown. He further showed that whether the larva that hatched from the egg resembled a diminutive adult (as it does in cockroaches, grasshoppers and the like) or whether it had a totally different form (such as the

42

caterpillar of a butterfly or the maggot of a fly) bore no relation to the amount of yolk the egg contained.

These ideas of Swammerdam are now generally accepted. The wide differences in form between the larva and the adult result from the independent evolution of larval and adult characters as these stages became adapted to totally different diets and modes of life; and the pupa probably arose as an intermediate form that would bridge the wide gap in morphological characters between larva and adult. This interpretation was put forward by Darwin and elaborated by Lubbock as the chief factor in the evolution of insect metamorphosis.

Swammerdam was unable to conceive the nature of these latent pupal and adult forms. We now recognize that they are latent in the gene system just like all the other characters of living organisms. The physiological study of metamorphosis consists in identifying the nature and action of the factors that regulate the change from one form to another.

The anatomical basis of metamorphosis: imaginal buds

The early discussions of insect metamorphosis centred around the more extreme examples, as seen in Lepidoptera, Hymenoptera or Diptera. The anatomical study of these forms led to the discovery of the ' imaginal discs ' which were observed by Lyonet (1762) in the goat moth caterpillar. The significance of these structures was first appreciated by Weismann (1864), in his classic work on *Calliphora*, as nests of embryonic cells set aside for the formation of the adult. They appeared to represent the latent imaginal organism in visible form; and physiological discussion turned upon what factors inhibit the development of these imaginal buds in the young stages, or activate the imaginal buds at metamorphosis and cause the larval tissues to disintegrate.

In most parts of most holometabolous insects (Lepidoptera, Coleoptera, etc.) and in all hemimetabolous forms, the adult cuticle is laid down by the same cells or by the daughter cells of those that have previously laid down the larval cuticle. Many parts of the body in Lepidoptera, which suffer a spectacular metamorphosis, are not formed from imaginal discs but are laid down by the same cells as have formed the larva. Towards the base of the legs in caterpillars there is a region of epidermis where cell division is very active during the last larval stage. In *Pieris* this region lies between segments two and three; it forms a ' differentiation centre ' from which the adult leg develops; it has commonly been referred to in the past as the imaginal disc for the adult leg. But in the previous larval stage this structure cannot be detected; at that time all the epidermal cells are concerned in laying down the cuticle of the larval leg.[216]

Imaginal discs thus exist in varying degrees in different insects.   In exopterygote insects, such as *Rhodnius*, all the epidermal cells are engaged in forming the visible cuticle throughout life.   But in the larva of endopterygote insects groups of epidermal cells become set aside for the formation of certain pupal or imaginal structures.   Early in life these cells may form part of the larval epidermis engaged in laying down larval cuticle.   Later they become invaginated as little epithelial pockets. Later still there is an evagination of epithelium outwards into the lumen of this pocket, and it is this evagination which forms the germ of the wing, leg or other organ.   In the limbs of Lepidoptera, as we have seen, the imaginal discs are not visibly distinguishable from the general epidermis until the end of the fourth instar.[218]   In such insects as *Musca*, *Calliphora* or *Drosophila* they are invaginated from the ectoderm towards the end of embryonic life or at a very early larval stage.   They become virtually detached from the larval epidermis, being connected only by a tenuous thread, the vestigial neck of the invagination.   And in addition to these compact embryonic buds there are in the higher Diptera numerous independent histoblasts or small groups of cells scattered throughout the body, notably beneath the larval epidermis. At the time of pupation (beneath the hardened larval skin, or puparium) the pupal cuticle covering the head and eyes and limbs is the product of the cells of the imaginal discs which are evaginated at this time.   But the cuticle covering the abdomen is laid down by the larval cells; not until after pupation are these replaced by a continuous layer of imaginal cells arising by proliferation from the groups of histoblasts.   In these higher Diptera, at the time the imaginal discs appear, all the epidermal cells are quite small; thereafter, those cells which form purely larval structures, the epidermis, fat body, etc., cease to multiply and grow only by increase in cell size; the small cells of the imaginal discs, on the other hand, remain small and the discs grow by multiplication of the cells.

It is clear from this brief review that the segregation of embryonic cells in the form of imaginal discs is not an essential feature of insect metamorphosis.   This arrangement is a peculiarity of certain holometabolous insects in which some structures show an excessive degree of growth at the time of metamorphosis.   By being invaginated from the epidermis and separated from it these cells are relieved of all responsibility for cuticle formation during larval life and are thus able to grow and multiply more or less continuously.

Metamorphosis consists of two processes: (*i*) changes in general form resulting from local outgrowths of the integument; and (*ii*) changes in cuticle structure resulting from changes in the activity of individual epidermal cells—coupled, of course, with related changes in the internal organs, muscles, nervous system, etc.   The imaginal discs are merely a special method of meeting the requirements of process (i).

Physiological control of metamorphosis: the corpus allatum in *Rhodnius*

Many hypotheses were advanced to explain the quiescence of the imaginal discs during the larval life of insects and their activation at the time of metamorphosis.   The explanation came ultimately not from the study of Holometabola but from experiments on Hemimetabola in which imaginal discs are absent and whose changes in form at the final moult were regarded by many entomologists as not being equivalent to metamorphosis at all.   It turned out that, like the process of moulting itself, metamorphosis is controlled by a hormone; but not, as might have been expected, by a ' metamorphosis hormone ' activating adult development, but by an anti-metamorphosis hormone which sustains larval development until the time for metamorphosis arrives.

The first indication of the nature of this control was obtained in *Rhodnius*.   We saw (p. 2) that *Rhodnius* larvae decapitated before the critical period after feeding failed to moult, and that during the critical period the proportion of decapitated larvae which moulted rose from 0 per cent to 100 per cent.   When larvae were decapitated during the second, third or fourth instars and the new cuticle of those which succeeded in moulting was examined, it was found that some of those which had been decapitated during the critical period showed a precocious metamorphosis.   They laid down cuticle with partially adult characters and showed partial differentiation of adult genitalia and wings. As can be seen in Fig. 14 the proportion of insects which showed this effect fell away rapidly; and the degree of differentiation of adult characters likewise became progressively less on successive days.   By the end of the critical period all the decapitated insects moulted and none showed any signs of metamorphosis.

These experiments clearly indicated that some influence exerted by the head in the early larval stages was preventing the realization of the imaginal characters latent within the cells.   This influence was shown to be hormonal in nature; for if a fifth-stage larva decapitated at 24 hours after feeding is joined in parabiosis to a fourth-stage larva (7 days after feeding) with only the tip of the head removed, it is not only caused to moult under the influence of the moulting hormone from the thoracic glands, but its metamorphosis is inhibited: instead of developing adult characters as would normally be expected, it again lays down larval cuticle.[211]

In the normal fifth-stage larva of *Rhodnius* the controlling hormone is no longer secreted, and metamorphosis takes place. This makes possible the converse of the experiment just described. If a first-stage larva of *Rhodnius*, recently hatched from the egg, is allowed to feed and is then decapitated one day later and joined in parabiosis to a fifth-stage larva in process of moulting, it will be caused to moult by the moulting hormone of the fifth-stage larva; but it will no longer be exposed to the anti-metamorphosis hormone and this newly hatched larva develops

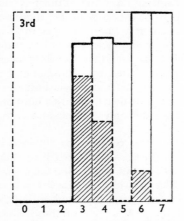 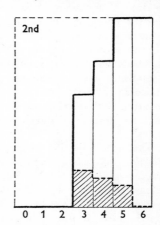

Fig. 14.   Charts showing the proportion of *Rhodnius* larvae in 3rd and 2nd stage which moulted, out of batches decapitated at the number of days after feeding indicated on the base line. The uppermost level represents 100 per cent moulting. The broken line (above shaded area) shows the proportion of insects developing adult characters.

directly into a diminutive adult, with adult cuticle and pattern, partially developed wings and quite well differentiated adult genitalia.[211]

The source of the hormone which prevents the appearance of adult characters in the young larval stages was readily traced to the corpus allatum: (*i*) removal of the anterior parts of the head containing the brain, but leaving the corpus allatum intact, did not lead to precocious metamorphosis; (*ii*) implantation of isolated corpora allata from the young larval stages (first, second, third or fourth) into fifth-stage larvae induced a further larval moult to giant, or sixth-stage, larvae (Plate ID); and when these larvae were fed they might moult again to form seventh-stage larvae[211] (Fig. 15).

The corpora allata were recognized by Nabert (1913), on structural grounds, as being probably ductless glands of unknown function.[212] The paired glands arise by the budding of ectodermal cells between the mandibular and maxillary segments of the embryo.   Later these cell nests become separated from the epidermis and form compact deeply staining bodies lying just behind the brain.   Sometimes, as in Hemiptera and the higher Diptera, they fuse to form a single median structure. They are innervated from the nerve to the corpus cardiacum (Fig. 25).

The corpus allatum hormone was at first called the 'inhibitory hormone'—inhibitory in the sense of maintaining the *status quo* and

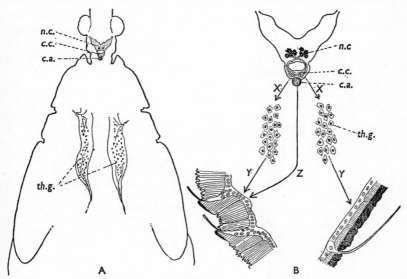

FIG. 15.   A, 5th stage larva of *Rhodnius* showing the location of the chief endocrine organs.   B, schema summarizing the hormonal control of metamorphosis.   *c.a.*, corpus allatum ;*c.c.*, corpus cardiacum; *n.c.*, neurosecretory cells; *th.g.*, thoracic gland.   X, activating hormone from the brain; Y, thoracic gland hormone (the moulting hormone 'ecdysone'; Z, corpus allatum hormone (the juvenile hormone).   To the left in B, larval abdominal cuticle is shown; to the right, adult abdominal cuticle.

preventing the realization of the latent imaginal characters.   It certainly has that effect; but when it was found later that the corpus allatum hormone would bring about a partial 'reversal of metamorphosis' (p. 57) it was regarded rather as actively favouring the differentiation of larval, or juvenile characters and since then has been called the 'juvenile hormone'[211] or 'neotenin'—the youth substance.

Control of metamorphosis in other insects

This general principle of the control of metamorphosis by secretion of juvenile hormone throughout the larval stages, and the cessation of secretion causing transformation to the adult, has been found to apply in almost all insects; but there are variations in detail that are characteristic of the different groups.   In the stick insect *Carausius*, which normally has seven larval instars before the female becomes mature, removal of the corpora allata at an early stage, say in the second-stage or third-stage larva, is followed by *two* moults and then the insect begins to lay eggs of half the normal size.[212]   Conversely the implantation of

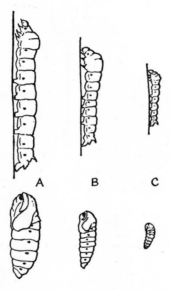

FIG. 16.   Precocious metamorphosis in the silkworm following removal of the corpora allata (after Bounhiol). A, normal 5th-stage larva and pupa; B, 4th-stage larva, and pupa resulting from ablation of corpora allata. C, 3rd-stage larva, and pupa likewise resulting from allatectomy.

corpora allata from young insects into *Carausius* in the fifth or sixth stage may cause them to make as many as four extra moults and become giant forms 15 cm long, or double their normal length.   A similar response is seen in the Madeira cockroach *Leucophaea*.   There are usually eight larval stages; if the corpora allata are removed in the fifth or sixth stage they make two moults before adultoid characters are developed.[212]   This delay presumably results from the persistence of substantial quantities of juvenile hormone in the blood and tissues of the young insects after the corpora allata have been removed.   In the cricket *Gryllus*,[212] the grasshopper *Melanoplus*[216] and in *Locusta*[216] extra larval stages can be produced by implantation of corpora allata.   In

Sialis (Neuroptera) precocious metamorphosis can be induced by decapitation or by extirpation of the corpora allata at any of the larval stages from the first to the ninth.[216]

Further variants are to be found in holometabolous insects. Bounhiol[14] early succeeded in extirpating the corpora allata in young silkworm larvae in the fourth, third and even in the second instar (Fig. 16). Such larvae then proceeded to spin cocoons and to pupate at the next moult; and the pupae gave rise to diminutive adults. The second-stage larvae produced pupae weighing only 25 mg as compared with the normal pupal weight of 1 g or more.[14] The tiny adults to which they give rise are more or less normal in appearance, and they lay eggs, not so large as usual, but by no means reduced in size in proportion to the body of the mother.[212] As in *Rhodnius*, so in all larvae of *Bombyx* after the first instar (as shown by Fukuda) metamorphosis can be induced simply by decapitation if this is done at the proper time—for example, in third instar larvae decapitated between 40 and 60 hours after moulting.[212] Similar results have been obtained in the wax moth *Galleria*; not only are diminutive pupae and imagines produced by removal of corpora allata from the very young larvae, but full-grown larvae receiving multiple implants of corpora allata from young larvae will continue larval development with the eventual production of giant pupae and adults.[212]

In *Tenebrio* (Coleoptera) as many as six extra larval moults have been obtained after implantation of corpora allata from young larvae.[212] The corpora allata have a controlling effect on metamorphosis in the honey-bee (Hymenoptera) which will be discussed later (p. 63). The major group in which the action of the juvenile hormone is not so clear is that of the Diptera. If the corpus allatum is taken from a young larva of *Calliphora* and implanted into the abdomen of a fifth-stage larva of *Rhodnius* a localized patch of larval cuticle is produced overlying the implanted gland;[212] and experiments by Vogt[212] and Possompès[216] on *Drosophila* and *Calliphora* respectively have afforded some evidence of the production of partially larval structures resulting from implantation of corpora allata into pupating larvae of these Diptera; but precocious metamorphosis, or giant larvae and adults, have not so far been produced. These are insects in which the imaginal discs are exceptionally important in adult development. Perhaps that is why such experiments have not proved successful. It might be expected that the more generalized group of Diptera, the Nematocera, might show such effects more readily. But the Tipulid *Nephrotoma suturalis* given massive doses of the juvenile hormone of cecropia (p. 69) by surface application or by injection into the pupa within a few minutes after the larval skin had been shed, developed into completely normal adults of both sexes.[220]

The other group in which the action of the corpus allatum is exceptional is the Apterygota, at the other end of the evolutionary scale. Here moulting continues in the adult insect, in which the thoracic glands persist; and the moults alternate with cycles of reproduction. In the silver fish *Lepisma* there is a constant correlation between the body size and the length of the genital appendages; the adult form is attained gradually. This relation is scarcely upset if the corpora allata are extirpated: it is not possible to produce adults of unduly small size as can be done in other insects.[216]

Control of juvenile hormone secretion

It is clear that metamorphosis is controlled by the corpus allatum. When this ceases to secrete the juvenile hormone, or secretes it in greatly reduced amount, metamorphosis takes place. But how is the activity of the corpus allatum controlled? In *Locusta*, metamorphosis seems to be decided in the first day after the fourth moult: if adult development is to be inhibited, active corpora allata must be implanted during the fourth stage or at the very beginning of the fifth stage. In the normal insect some factor inhibits the activity of the corpus allatum at the beginning of the fifth stage. L. Joly[216] postulated some unknown organ secreting an inhibitor substance. But in *Rhodnius*, as we have already seen, the isolated corpus allatum of a fourth-stage larva, when implanted into a fifth-stage larva, will continue to secrete the juvenile hormone so that its host moults to a sixth-stage larva. The implanted fourth-stage corpus allatum will now be at the fifth-stage; but the host may moult again to a seventh-stage larva in which very little further differentiation of adult characters has taken place. Clearly the corpus allatum is still secreting juvenile hormone. It was, therefore, inferred that secretion is normally inhibited by nervous control from the brain.[211]

It is evident that the corpus allatum cannot be wholly autonomous; it does not, for instance, produce juvenile hormone for a predetermined number of moulting stages. It is not the corpus allatum which ' counts the instars '. The mechanism responsible for this presumably resides in the central nervous system and the requisite stimuli must be conveyed to the corpus allatum by the nerves with which it is richly supplied. The nature of the nervous control is not fully understood, but from a survey of the literature Scharrer[216] was led to suggest that the restraining influence on the corpus allatum is nervous, the stimulating effect is by some substance in the neurosecretory product from the brain.

If the corpus allatum is removed from the fifth-stage larva of *Pyrrhocoris* 5–6 days after moulting, and is implanted into the

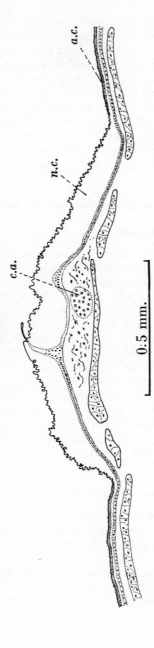

Fig. 17. Section through a localized patch of larval cuticle in an otherwise normal adult *Rhodnius*. *c.a.* implanted corpus allatum; *a.c.* normal adult cuticle, thin with relatively thick exocuticle; *n.c.* larval cuticle, very thick and soft with highly folded epicuticle.

newly moulted fifth-stage larva, a number of these develop partial or complete larval characters when they moult; and the same happens in *Galleria* with corpora allata removed from larvae a day or two before pupation.[216]   These results could be explained by the release of the corpus allatum from inhibition by the brain when the nerve supply is severed.   Likewise in *Bombyx*, the isolated corpus allatum retains its activity and may, indeed, regain its activity, even when completely isolated from the brain.   This would agree with the existence of a nervous inhibition by the brain, but it does not support the view that the secretion of the juvenile hormone is dependent upon the supply of neurosecretory products from the brain.[58]

The action of the juvenile hormone at the cellular level

Local action of the juvenile hormone

When the corpus allatum of a young insect is implanted into a larva during the last instar, instead of turning into a giant larva the new host may sometimes become an adult with no more than a local patch of larval cuticle overlying the implanted gland (Plate IIB); and this patch may often show a sharply defined boundary (Fig. 17).   Clearly the hormone is acting directly upon the epidermal cells.   The same result can be obtained by the use of active extracts (p. 68) containing juvenile hormone: by applying such extracts to the surface of the cuticle in restricted parts of the body, it is possible to produce adults with one larval tergite in the abdomen; with one larval wing; or with larval genitalia in an otherwise normal adult (Plate II).

In all these experiments the juvenile hormone seems to be merely exerting an inhibitory influence on the appearance of adult characters. It was for this reason that it was at first called the ' inhibitory hormone ' or, by Williams the ' status quo ' hormone.

Juvenile hormone action as an arrest of ' differentiation '

Indeed, the first attempt to explain the action of the juvenile hormone was to picture metamorphosis as a progressive differentiation.   It was supposed that two competing processes were at work: (*i*) the differentiation in structure towards the adult form; (*ii*) the secretion and deposition of the new cuticle.   If the deposition of cuticle were to be accelerated, there would be less time available for adult differentiation and a partial inhibition of metamorphosis would result[211] (Fig. 18).   This idea was derived from the conception of Goldschmidt that the relative rates of different growth processes are responsible for controlling body form, pigment patterns and the like.

Small degrees of progressive enlargement, or ' differentiation ', in the genitalia and wing lobes take place during the five larval moults in *Rhodnius*; and it was shown by experiment that if a third-stage larva, after the critical period, with the corpus allatum intact, was joined in parabiosis with a fourth-stage larva at one day after feeding, then the fourth-stage larva when it moulted was found to have developed

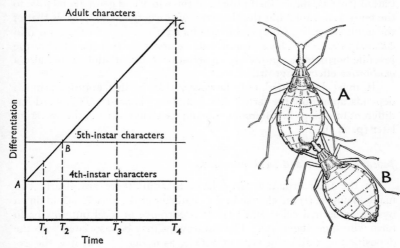

FIG. 18.   To the left, diagram to illustrate the way in which differentiation toward the adult form can be arrested at different points by accelerated deposition of the new cuticle at times T_1, T_2 etc.   To the right, accelerated cuticle deposition induced in the 4th stage larva (A) by joining to it a larva (B) already at an advanced stage in moulting.

characters intermediate between a fourth-stage larva and a fifth.   This partial arrest in differentiation was believed to be an effect of the corpus allatum of the third-stage larva.   But the same thing happens to a fourth-stage larva (at one day after feeding) if it is put in parabiotic combination with another fourth-stage larva at seven or eight days after feeding (Fig. 18).[212]

This result is certainly due to partial suppression of differentiation by accelerated deposition of the new cuticle; but this speeding up is brought about not by the juvenile hormone but by the moulting hormone ecdysone.   The same result was obtained by Halbwachs, Joly and Joly[216] by implanting several active ventral glands (secreting the moulting hormone) into a fifth-stage larva of *Locusta*; and it can be reproduced in *Rhodnius* by the injection of excessive amounts of ecdysone. Injection of 5 μg of ecdysone into a fourth-stage larva of *Rhodnius* (the normal requirement, as we have seen, is in the order of 1 μg) causes it to develop characters intermediate between those of a fourth- and fifth-instar larva. An excessive injection of 5–10 μg into a fifth-stage larva results in an adult with small and incompletely differentiated wings and genitalia. This effect differs from that of the juvenile hormone in a very important respect: there is indeed reduced differentiation (that is, diminished growth in the wings and genitalia relative to the body as a whole) but the character of the cuticle laid down is that of the adult stage.[220]   Comparable results were obtained in the Saturniid *Samia cynthia*.[224]   These experiments demonstrate that the effect of the juvenile hormone is not merely an arrest of differentiation; it has also a *qualitative* effect on growth.

It must be pointed out, however, that in this argument much depends upon how the term ' differentiation ' is used. This word has different meanings in different contexts; we shall return to this question later (p. 113).

Juvenile hormone and local outgrowths

In the transformation from larva to adult there are clearly two major changes: (*i*) the change in the character of the cuticle as laid down by the epidermal cells; and (*ii*) the occurrence of local outgrowths to form wings, genitalia, etc. The outgrowths may be associated with the deposition of a different type of cuticle, as under (*i*); but it is the exaggerated local growth that is commonly the more striking element in metamorphosis. As we have already seen (p. 44) such outgrowths may result from excessive enlargement of larval structures which already exist: as in the wings, the prothorax and genitalia in *Rhodnius*, or in the legs of Lepidoptera; or they may result from the growth and evagination of the imaginal discs. Thus absence of the juvenile hormone permits active growth of the imaginal discs.

In these circumstances, when attention is focussed on the local outgrowths, the first sign of metamorphosis (that is, of lack of juvenile hormone) is local mitosis. In the genitalia and wing lobes of the fifth-stage larva of *Rhodnius*, mitoses are plentiful within 48 hours of feeding; whereas in the presence of juvenile hormone mitosis in these parts is relatively restricted and does not appear until about six days after feeding.

But more active mitosis is not necessarily associated with *lack* of juvenile hormone. In the epidermis over the abdomen of the *Rhodnius* larva the cells are closely packed and the cuticle highly folded, to allow distension of the abdomen to receive the large meal of blood. In the adult *Rhodnius*, where the greater part of the abdominal cuticle is incapable of being stretched, the epidermal cells are much more sparse. It follows, therefore, that in these regions of the body mitosis is much more active in the *presence* of the juvenile hormone.

Larval and adult body 'patterns'

It was implied (p. 44) that there are two types of cuticle: larval cuticle and adult cuticle; and that metamorphosis represents a switch

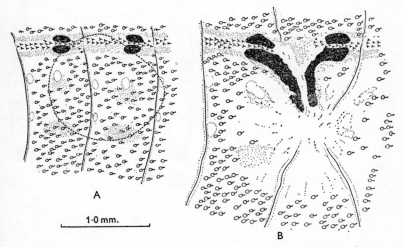

A

1·0 mm.

B

FIG. 19. A, part of two segments of the abdomen of a 3rd-stage larva of *Rhodnius* showing the black pigment spots at the margin between tergites and sternites. The broken line shows the area where the epidermis was killed by burning. B, the corresponding region after healing and moulting to the 4th stage.

from the one to the other. But there are, in fact, many types of cuticle in both larva and adult, varying in structure and composition, in surface sculpturing, in pigmentation, in the incidence and character of the hairs present and so forth. Indeed, the larval stages and the adult stage each have a characteristic 'pattern' visible in the cuticle. It would, therefore, be more correct to say that metamorphosis represents a switch-over from the larval pattern to the adult pattern. These two patterns are both carried separately by the same epidermal cells. At a given moment the one pattern is visible in the cuticle that has been laid down, the other pattern is latent in the cells and invisible.

IH C

This can be strikingly demonstrated by some old experiments on *Rhodnius*.[211] The special characters of the epidermal pattern may be retained by the epidermal cells when they multiply and spread during the healing of a wound. If a burn inflicted on the abdomen in the *Rhodnius* fourth-stage larva, passes through the marginal black spots on the tergites, the wound is repaired by means of cells spreading inwards, from the margin of the burn (p. 117). There is, therefore, a migration inwards and repeated division, of cells derived from the region of the

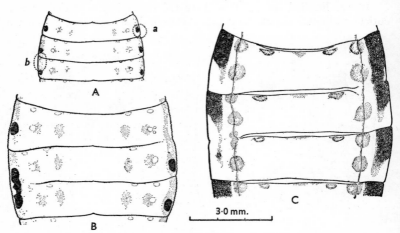

FIG. 20. A, third, fourth and fifth tergites of a ' normal ' 3rd-stage larva of *Rhodnius*. The broken lines at *a* and *b* show the regions burned. B, corresponding segments in the 5th-stage larva resulting. C, corresponding segments in the adult resulting.

black spots. These daughter cells carry with them the potentiality to lay down black cuticle; when the new cuticle of the fifth-stage larva is formed it therefore shows a centripetal displacement of the black patches (Fig. 19B).

If a small burn is applied between two black spots on the lateral hind margin of successive segments in the third-stage larva (Fig. 20A), these are found to have joined up when the insect moults (Fig. 20B); if the burn includes the whole of one black spot (Fig. 20A) it is eliminated and is absent from the resulting fifth-stage larva (Fig. 20B). But the black spots which appear in the adult *Rhodnius* lie at the anterior angles of each tergite, and the posterior angles are pale. Therefore, when the insect just described undergoes metamorphosis at the moulting of the fifth-stage larva, the effects on the pattern are reversed. Where the black spots of the larva had become united the adult black spot has been eliminated, and where the black spot of the larva that had been eliminated, the black spots of the adult have become fused (Fig. 20C).

There is no question here of progressive differentiation towards the adult form. There are two forms, or patterns, both equally specialized; and the one can clearly remain latent in cells which are engaged in producing the other. The problem of metamorphosis is the problem of the control of these alternative patterns.

Localized mitosis as an element in the body pattern

In a paper on ' homeostasis and insect growth '[217] the idea was developed that a capacity for localized mitosis and outgrowth is just as much an element in the body pattern as is a localized capacity for laying down black pigmented cuticle; and that localized mitosis (coupled with an appropriate mutual relation between the cells, that controls the body form) will lead to all the spectacular changes in form that characterize metamorphosis. It is not necessary, therefore, to picture two adult patterns, one concerned with outgrowths and one concerned with cuticle structure. It is sufficient to picture a single pattern embracing these two elements.

What is the nature of the invisible adult pattern? It is clearly a pattern of interlocking ' fields ' in each of which the cells are already ' determined ' to form a particular component in the pattern of the body. As we have seen, when the cells divide in the presence of the juvenile hormone, the adult pattern remains latent and invisible; but it is carried and maintained by the daughter cells. That suggests, though it does not prove, that the elements of the pattern are carried by the chromosomes; that what the juvenile hormone is doing is ensuring the activity of the gene pattern characteristic of the larva; and that metamorphosis is a genetic switch, brought about by the absence of the juvenile hormone, which affects every element in the body pattern.[217]

That may be taken as the shorthand description of the action of the juvenile hormone at the cellular level. It is not known whether it acts directly upon the nuclear genes or whether it acts primarily on the cytoplasm. The name ' juvenile hormone ' has sometimes created the belief that it is an anti-ageing factor. There is, however, no evidence for any action of that kind; it is merely a morphogenetic hormone—one of the best examples of a really clear-cut morphogenetic hormone to be discovered.

The juvenile hormone and the 'reversal' of metamorphosis

The adult *Rhodnius* can be induced to moult again by joining it in parabiosis with one or more fifth-stage larvae that have passed the critical period, or by implanting active prothoracic glands or by injection of ecdysone (p. 12). Under these circumstances it produces a new cuticle of wholly adult type. But if, at the same time, six or so active corpora allata from a moulting fourth-stage larva are implanted into the abdomen, or if an active preparation of juvenile hormone (p. 68) is injected, the new cuticle formed when the adult moults shows a partial reversion to larval characters. This change does not affect elaborately differentiated structures such as the genitalia, but it is evident in the cuticle structure and pattern of the abdomen.[211]

Results identical in principle were obtained by Piepho[212] in the wax moth *Galleria*: when fragments of pupal integument were implanted into young larvae and were thus induced to moult in a larval environment, they laid down a cuticle of larval type. In other experiments Piepho and his colleagues implanted fragments of the integument of full-grown larvae of *Galleria* into the abdomen of other full-grown larvae. They duly formed pupal cuticle, followed by adult cuticle, synchronously with their host. They were then removed and implanted into young larvae, pupating larvae, or into pupae, and finally examined in section when this second host had become adult. It was found, as in *Rhodnius*, that the imaginal integument would readily moult again to produce an adult cuticle with scales of almost normal form. In a smaller number of experiments it had been possible to induce it to form pupal cuticle, and when the host moulted to become adult the implant again became of imaginal type with scales. In this experiment metamorphosis was first taken backwards and then forwards again.[212]

Two other examples of reversal of metamorphosis may be quoted. Ozeki[216] showed that if adult earwigs (*Anisolabis*) retaining their own corpora allata (and therefore providing their own juvenile hormone (p. 76)) were caused to moult by the implantation of active ventral glands, they developed once more the ecdysial line on the thorax that is characteristic of the larval stages. This reversal of metamorphosis does not occur if the moulting adults have been deprived of the source of juvenile hormone by removing their corpora allata.

Lawrence[106] transplanted fragments of the epidermis and cuticle from adult milkweed bugs (*Oncopeltus*) into the abdominal wall of young (third-stage) larvae. In the course of two moults in the presence of juvenile hormone, these implanted fragments were again producing cuticle of partially larval character. Once again there has been a partial reversal of metamorphosis; but, as in the other insects studied, the cells are much more reluctant to make this backward movement in morphogenesis than they are to make the normal change of metamorphosis to the adult. We shall return to the consideration of this phenomenon in the next section (p. 61).

The action of the juvenile hormone on gene activity in the cell

As has been emphasized already, the epidermis of the insect is particularly favourable material for the study of differentiative changes during growth because it expresses its juvenile or adult character by the type of cuticle it lays down.   It may therefore be possible to judge whether the activity of a single cell is larval, adult or intermediate.

The cuticle has a complex structure, the product of a complex system of enzymes within the cell, its characters determined by the nature of this enzyme system.   Since a given cell may lay down soft pale cuticle in the larva and hard dark cuticle in the adult, there must be large changes in the enzymes that are acting in the two stages.   But, as we have seen already, the differences are equally great as between one region of the body and another; a hormone which is leading to the production of soft pale cuticle in place of hard black cuticle in one part of the body, and hard black cuticle in place of soft pale cuticle in another part, must be acting in some very general way.   And since single enzymes are believed to be products of single ' genes ' it is concluded that the absence of the juvenile hormone is commonly concerned in switching the gene system throughout the body from producing larval enzymes to producing adult enzymes.

If the juvenile hormone is switching gene activity, it might be expected that, confining our attention to a single type of cell, it would be possible to observe intermediate stages in which certain activities, that is, certain enzymes or groups of enzymes, had been changed while others had not; so that the cuticle produced might be adult in some respects but still larval in others.

The trichogen cells which form the cuticular hairs persist from one instar to the next.   If fourth-stage larvae of *Rhodnius* are decapitated around the critical period (Fig. 21) it is possible to select, from among those larvae that moult, a graded series with cuticular hairs showing all degrees of intermediate characters between larval hairs and adult hairs.[211]

Conversely it is possible to observe a similar step-like process of switching if fifth-stage larvae of *Rhodnius* are allowed to develop normally towards the adult form for a known number of days after feeding, and are then switched back to larval development by flooding the system with juvenile hormone.[211]   Larvae switched at one particular phase of development have lost the capacity to produce smooth unpigmented larval plaques on the abdomen, but they show spots of wrinkled unpigmented cuticle marking the places where plaques would have appeared if the switch back had been made a little earlier.   Thus the ability to form smooth weakly sclerotized cuticle is lost by the cells below the plaques before the ability to form unpigmented cuticle. Larvae which are switched back a little later show uniformly grey wrinkled cuticle; the capacity to form unpigmented spots has now been lost.

In the larva of *Oncopeltus* in which, as in many insects, the area of cuticle laid down by each epidermal cell has clearly visible boundaries, it has been possible to show very elegantly the progressive changes in the type of cuticle laid down by single cells as this process of switching goes forward.[106]

There is another aspect of the control of metamorphosis which is

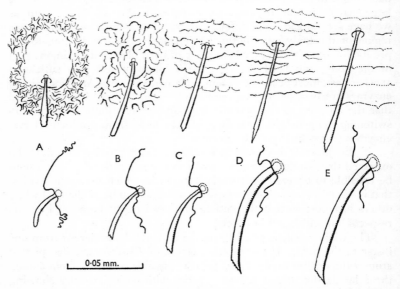

FIG. 21. Precocious metamorphosis and intermediate stages in the sensory hairs of *Rhodnius*, produced by decapitation of the 3rd-stage larva (see fig. 14). *a*, normal 4th-stage larva; *b, c* and *d*, grades of metamorphosis in hairs formed by decapitated larvae; *e*, normal adult hair for comparison. (Above, hairs on surface of abdomen; below, from margin of abdomen).

relevant to this discussion. The cells seem to acquire a certain 'inertia ', such that once they have begun to form pupal or adult structures they resist the action of the juvenile hormone in causing them to revert to the larval state. Presumably the cell has been ' re-programmed ' and those components of the gene system (or those segments of the DNA chain in the chromosomes) that are responsible for the production of the adult enzymes, have been brought into action, and this process is not easily reversed.

It was shown long ago by Piepho[212] that this inertia was soon eliminated in the regenerating cells responsible for wound repair.   Cell division favours the changing over of the gene programme, presumably through the synthesis of *new* DNA.   On the other hand, there are undoubted examples, as outlined above, of single cells which undergo neither cell division nor endomitosis, in which the programme of enzyme synthesis can be changed.   The question arises (*i*) whether there exist in the cytoplasm of such cells preformed templates (sometimes referred to as persistent or long-lived messenger-RNA) which can be brought into action for the synthesis of a changed type of cuticle by a change in the hormones—that is, by the presence or absence of the juvenile hormone; or (*ii*) whether the juvenile hormone acts directly on the DNA of the nucleus and evokes a change in the activity of the gene system.

Whatever may be the actual site of action of the juvenile hormone within the cells, the fact that the cells can be caused to undergo metamorphosis and then to revert to laying down larval cuticle, fits in well with the idea that the programming of the gene system can be switched in either direction.   The cells are certainly far more resistant to switching their activity from the adult state to the larval state than they are to switching from larva to adult; but perhaps that is not surprising since the pressure of selection will naturally favour the normal progress of metamorphosis.

The effects of juvenile hormone on internal organs and on behaviour

In Hemimetabola there are usually no striking changes in most of the internal organs at metamorphosis; but the gonads, which are essentially adult structures, suffer the same suppression of growth and differentiation during the larval stages as do the external genitalia.   If precocious metamorphosis is induced in the third-stage larva of *Rhodnius* by parabiosis with a fifth-stage larva, the ovaries undergo a partial differentiation towards the adult state.[211]   The same happens in the cockroach *Periplaneta*[212] and in other insects; but in all cases the gonads fail to develop fully adult characters if they are taken from a larva at too young a stage.   They must reach a certain size before they are ' competent ' to undergo the change of metamorphosis.   Perhaps ' competence ' merely means the presence of a sufficiently large number of cells.

When observing the effect of the juvenile hormone in restraining the differentiation of the gonads until the time of metamorphosis, it is impossible not to be reminded of the metamorphosis that occurs in man at puberty;[216] and that parallel at once suggests the possibility that in the evolution of insects the earliest function of the corpus allatum hormone may have been to delay the differentiation of the gonads until growth elsewhere in the body was sufficiently advanced; and that the control of all the other characters associated with metamorphosis is a secondary effect that was acquired later.   We saw earlier that exposure of the tissues to juvenile hormone at different times during the metamorphic moult will produce different intermediate effects on the external pattern of the cuticle (p. 59).   If this same type of experiment is performed during the last larval stage of *Galleria*, various intermediate forms of the internal organs can likewise be produced.[178]

An interesting example of a partial reversal of metamorphosis is afforded by the alimentary canal of *Galleria*.   If the atrophied mid-gut of the adult wax moth is implanted into the body cavity of the caterpillar, when this moults again the basal cells of the gut wall begin to grow and multiply once more and develop a mid-gut of larval type.[216]

The juvenile hormone may also exert effects upon behaviour. Larvae of *Galleria* spin a slight flat web of silk on which they settle down before they make a larval moult; but before moulting to the pupa the full-grown larva spins a tough cocoon.   As was shown by Piepho, if active corpora allata are implanted into the last stage larva, so that it moults again into a larva, it will spin a larval type of web or an intermediate type.[216] In the Sphingid *Mimas tiliae*, if the juvenile hormone is above a threshold concentration the larval behaviour is that which is normal before a larval moult.   If the hormone is below this threshold the larva migrates down the tree trunk to the soil, as before a pupal moult.[216] It may well be that these changes in behaviour are the sequel to changes in growth and development within the central nervous system which are under the control of the juvenile hormone.   In *Pyrrhocoris apterus* large doses of substances with juvenile hormone activity (p. 67) cause complete retention of larval morphology and physiology; intermediate doses cause formation of larvae with perfect larval external morphology but with some adult patterns of sexual behaviour, degeneration of thoracic glands and appearance of diapause; and low doses cause formation of adultoids with larval structures more or less preserved but with adult internal anatomy and behaviour.[229]

Control of the steps towards metamorphosis

In Hemimetabola, such as *Rhodnius* and *Locusta*, with a set number of larval stages, each instar has its characteristic morphology.   This is evident chiefly in the progressive differentiation of the wing lobes and

the genitalia.   As already pointed out these changes seem to be regulated
(under the continuing action of the juvenile hormone) by the level of
ecdysone secretion, which controls the time available for further differentiation before these changes are brought to an end by the deposition of the new cuticle (p. 53).   In Holometabola there is a strikingly
distinct pupal stage interposed between larva and adult.   The question
arises as to how the development of the pupa is controlled.

The early work of Bounhiol[14] on the control of metamorphosis in
the silkworm *Bombyx* showed that removal of the corpus allatum in the

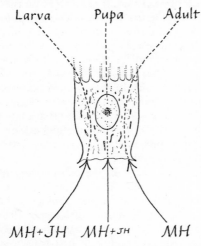

Larva   Pupa   Adult

FIG. 22.   Hormonal con-
trol of metamorphosis in
epidermal cell acted upon
by   moulting   hormone
(MH) and juvenile hor-
mone (JH) in different
proportions.

MH+JH   MH+JH   MH

young larva led to immediate transformation to the pupa.   He concluded
that metamorphosis in Lepidoptera takes place at the change from larva
to pupa.   He made no mention of any intermediate forms between larva
and pupa or between pupa and adult.   On the other hand, when
Piepho[212] implanted a variable number of corpora allata into the last
stage larva or the pupa of *Galleria*, or when the implants were made on
progressively later days, he was able to obtain a graded series of intermediates between larva and pupa and between pupa and imago.   He
concluded that before the larval moult there is much juvenile hormone
in the body, before the pupal moult much less, and before the imaginal
moult less still (Fig. 22).

Conversely, if the corpus allatum is eliminated by decapitation in
the honey-bee larva at the appropriate time, the normal pupal stage may
be omitted and incomplete imaginal characters develop when the larva
moults.[212]   Patches of larval integument implanted into the body wall
of the pupa in *Galleria* and other Lepidoptera will moult directly to

produce a cuticle with scales or structures intermediate between setae and scales, the normal pupal stage having been omitted;[212] and if the corpora allata are excised from the last stage larva of *Hyalophora cecropia* the resulting pupa (as in the honey-bee) shows partially adult characters.[216] Finally, it has been shown that corpora allata taken from larvae of *Bombyx*[216] or larvae of *Hyalophora*[216] become a very weak source of juvenile hormone towards the time of pupation.

These results fit in well with the conclusion of Piepho that the larval characters are determined by high concentration of juvenile hormone; pupal characters by a low concentration; adult characters by the complete absence of the hormone. But the fact remains that in Lepidoptera the cuticle, or patches of cuticle, laid down in experimental animals are nearly always larval, or pupal, or imaginal. Intermediates are, on the whole, uncommon; and when corpora allata are excised from a caterpillar, no matter what instar it may be, it always forms a pupa, which develops into the imago. It seems unlikely that the concentration of juvenile hormone remaining in the blood and tissues after such operations should always be the same. It would seem rather that there must be some mechanism tending to canalize development along one or other of three lines. That mechanism is presumably the selective activation of preformed alternatives in the gene system.

The same surprising constancy is seen in the morphology of the sixth-stage larva of *Rhodnius* (Plate ID). This instar never occurs naturally, but when produced experimentally its morphology is just as characteristic and as constant as that of the normal fifth instar (Plate IB) —no matter whether it is produced by the implantation of a corpus allatum of the fourth-stage *Rhodnius*, a number of first-stage corpora allata, the corpus allatum from *Periplaneta*, or the application of various chemicals with juvenile hormone activity (p. 67).

In many caterpillars there are striking changes in characters, such as pigmentation or the form of the setae, etc., in each successive instar. Attempts to prove that these changes are controlled by different levels of juvenile hormone secretion have given equivocal results. Certain insects undergo what is commonly called a ' hypermetamorphosis '. In Meloidae the actively feeding larva of typical caraboid form undergoes regressive changes, with shortening of the appendages and rounding of the body, and enters a resting stage or pseudochrysalis, often hardened like a puparium. It then emerges once more as a motile larva, now of scarabaeoid form, which in course of time moults into the true pupa. Nothing is known about the hormonal regulation of these changes either. Animal development follows a set pattern or programme. To what extent the successive stages in development are obligatory manifestations of this programme; or to what extent they are subject to modification by hormones or other factors, can only be decided by experiment.

Hormonal balance in development: prothetely and metathetely

The moulting hormone and the juvenile hormone are sometimes described as being antagonistic. In fact, they do two quite different things. Ecdysone brings about the activation of the epidermal cells that is necessary for growth and moulting. Once growth has been set going in this way the production of larval or of adult characters is determined by the presence or absence of the juvenile hormone. On the other hand, if too much ecdysone is present (p. 52) cuticle deposition may go forward so rapidly that insufficient time is available for the full development of adult outgrowths or for the breakdown of cells producing larval structures; hence intermediate forms may be produced (p. 53). Conversely if the juvenile hormone is secreted in insufficient quantity for the production of the larval form, there may be a partial development of adult characters; again intermediate forms may appear (p. 46). The lateral neurosecretory cells of the protocerebrum in *Locusta* normally excite the activity of the corpus allatum in secreting juvenile hormone. Hence if these cells are destroyed by electric cautery in the fourth-stage larva, dwarf adults are produced. But the neurosecretory cells of the medial region of the pars intercerebralis have the converse effect of inhibiting secretion by the corpus allatum: if excess of these cells are implanted, dwarf adults may again be produced. The control of corpus allatum activity is effected by the balance of these two effects.[67]

Long before the hormonal regulation of metamorphosis was recognized abnormal insects had been described in which larvae had developed partially pupal characters—such as silkworms developing enlarged antennae, or mealworm larvae developing wing pads. This precocious appearance of imaginal characters was termed 'prothetely'. Other examples were reported in which adult insects had retained some pupal characters, or pupae had retained some larval characters. This persistence of juvenile characters was termed ' metathetely '; it is equivalent to ' neoteny '.

When the hormonal regulation of metamorphosis was defined it was pointed out that these anomalies are most readily explained as resulting from an upset in the balance or the timing of hormone secretions.[211] This upset may be induced by exposure of cultures of *Tenebrio* or *Tribolium* to high temperatures or to low temperatures. They sometimes occur in larvae of *Simulium* invaded by the nematode *Mermis*. In Lepidoptera they are particularly liable to occur in species hybrids—

again presumably because the hormone balance has been upset.[212] Partial retention of larval characters in Saturniid pupae and in *Tribolium* maintained in culture has been traced to infection by *Nosema* and is attributed to the liberation by the micro-organism of substances with juvenile hormone activity.[54]

Intermediate forms may also be produced in insects exposed to various poisons which interfere with differentiation. When *Tenebrio* pupae are injected with Actinomycin D, Mitomycin C, barbiturate, etc., they may develop normal adult legs, wings and thorax, but a partially pupal abdomen. Imaginal development has already been determined in the head and thorax by the time of injection and these are, therefore, unaffected. But determination in the abdomen occurs later and the switch over from pupal to adult development is partially arrested by these metabolic poisons.[33]

Juvenile hormone and the control of cell death

Embryonic growth is always associated with the production of more cells than are necessary for development. The unwanted cells die and undergo autolysis. During the changes in form that occur in post-embryonic growth whole tissues and organs may become surplus to requirements and they likewise undergo autolysis. Autolysis of cells combined with ' chromatolysis ', the formation of ' chromatin droplets ' derived from the dissolution of excess nuclei, is particularly evident at metamorphosis when the cells responsible for purely larval structures no longer required are breaking down. Most if not all of these chromatic droplets are the product of whole nuclei that have disintegrated.[218]

The juvenile hormone plays an important role in maintaining the viability of the various specialized epidermal cells responsible for certain of the cuticular structures of the larva. During the larval stages in *Rhodnius* the hair-forming trichogen cells throughout the abdomen persist from one instar to the next. When the fifth-stage larva undergoes metamorphosis to the adult, hairs are no longer developed over the greater part of the tergites and the trichogen cells break down. The juvenile hormone prevents this break-down.

An important effect of the juvenile hormone is the maintenance of the prothoracic glands which will enable the larval insect to undergo a further moult. As we have seen (p. 9) these glands suffer autolysis in *Rhodnius* within 24 hours after moulting to the adult and within 48 hours the nuclei have disappeared completely. In other insects the glands break down at varying times before or after the adult moult. If supernumerary larval stages are produced experimentally by the implantation of active corpora allata, they retain their prothoracic glands and are therefore able to moult again.[212]   If the prothoracic gland is removed from an adult *Rhodnius* immediately after moulting it will break down in any environment into which it is introduced, even in a fourth-stage larva with juvenile hormone present.   On the other hand if the gland is removed from the fifth-stage larva shortly *before* moulting and is implanted into a young adult it undergoes no breakdown.   It appears, therefore, that when the prothoracic gland goes through a secretory cycle in the absence of the juvenile hormone, it will break down; but that some further humoral factor, to which the gland is exposed at or immediately after moulting to the adult, is needed to precipitate this breakdown.[213]   The nature and source of this second factor are unknown.

Histology of the corpus allatum

The active corpus allatum in *Rhodnius* consists of closely packed cells with a relatively large amount of homogeneous cytoplasm; the inactive gland is often smaller, but the main change is in the shrinkage of the cytoplasm with the appearance of vacuolated spaces between the cells (Fig. 24).[211]   In *Leucophaea* the active gland likewise shows increased cytoplasm with separation of the nuclei; with the cessation of activity during pregnancy (p. 79) the cell number falls and pycnotic nuclei can be seen.[216]

As seen in sections under the electron microscope, the inhibited glands show closely packed nuclei with clumped chromatin, and little cytoplasm, with deep infolding of the cell boundaries.   These stellate membranes straighten out as the cytoplasm expands during secretory activity; the nuclei enlarge and the mitochondria and ribosomes in the cytoplasm increase.[216]   The most striking feature of the cytoplasm of the corpora allata cells in *Locusta*,[90] in *Schistocerca*[147] and in Blattids[216] is the abundant network of smooth endoplasmic reticulum, recalling the steroid secreting cells of vertebrates.

Chemistry of the juvenile hormone

The juvenile hormone is non-specific: parabiosis between *Rhodnius* and *Triatoma*, or *Rhodnius* and *Cimex*, showed very early that the hormone would act on insects belonging to different families.[211]   It has been found that it is equally effective over a far wider range.   The corpus allatum of adult *Periplaneta* implanted into the Lygaeid *Oncopeltus* or into *Rhodnius* will lead to the production of giant larvae.   We saw (p. 49) that the ring gland of *Calliphora* larvae will cause local formation of larval cuticle in *Rhodnius*.   By using the full-grown larva of

Galleria as a test insect Piepho has shown that transplanted corpora allata from *Ephestia*, *Bombyx*, *Tenebrio* and *Carausius* are all equally effective.[212]

Active extracts of the hormone were first obtained by Williams[216] from the abdomen of the adult male of *Hyalophora cecropia* and this still remains the most active source known. Simple extraction of the abdominal lipids with ether gives a deep orange coloured oil which shows strong juvenile hormone activity. Similar extracts from other insects usually have demonstrable activity only after suitable fractionation. When active extracts had been obtained it became possible to devise simple methods of assay which would serve to follow the active principle during fractionation procedures, and thus to demonstrate the presence of material with juvenile hormone activity from other sources.

In the course of such assays it was found that the active substance would penetrate the cuticle; although, when the concentration was low, as in the crude extract from *Hyalophora*, it might be necessary to abrade the surface, as in *Rhodnius*, or to make punctures in it, as in the pupa of *Tenebrio*.[216] It was also found that the juvenile hormone was readily broken down in metabolism and that this breakdown was greatly accelerated if the oil solution was emulsified before injection.[66] It was most effective when the active principle was continuously liberated in small amounts from a store. That could be achieved by the injection of a drop of oil containing a dilute solution[66, 219] or by applying the substance mixed in the paraffin wax used to seal a small wound in the integument: the ' wax test '.[66] A deposit of the active substance on the surface of the cuticle can likewise serve as a store which slowly releases the hormone into the insect.

By the use of such assays it was possible to work out methods for the rapid concentration of the active principle in extracts. These methods were based chiefly on counter-current separation in suitable solvents by means of which the active substance was readily concentrated along with the non-saponifiable fraction in the oily extracts. Material with juvenile hormone activity was thus extracted from many insects and from many other invertebrates.[66] It was demonstrated in the suprarenal glands of cattle and in many other organs of vertebrates, including man[216] and even in some plants, bacteria and yeasts.[216] Indeed Saturniid larvae[54] and *Tribolium* heavily infected with the microsperidian parasite *Nosema* will show juvenile hormone effects at metamorphosis.

Active material was first isolated in pure form from the excreta of *Tenebrio* and from yeast by Schmialek[216] and was shown to be a mixture of the open chain terpene alcohol farnesol (I) and its aldehyde farnesal (Fig 23). Farnesol is a very widespread metabolite, an intermediate in the biosynthesis of cholesterol and of carotinoids; although it had rather low juvenile hormone activity in insects these results at once suggested that the natural hormone might well be a related isoprenoid compound. Many such substances have been synthesized. Schmialek[216] found that if the alcohol farnesol was converted into the methyl ether (II) it was much less readily broken down in the insect (*Tenebrio*) and was far more active. Farnesyl methyl ether proved highly

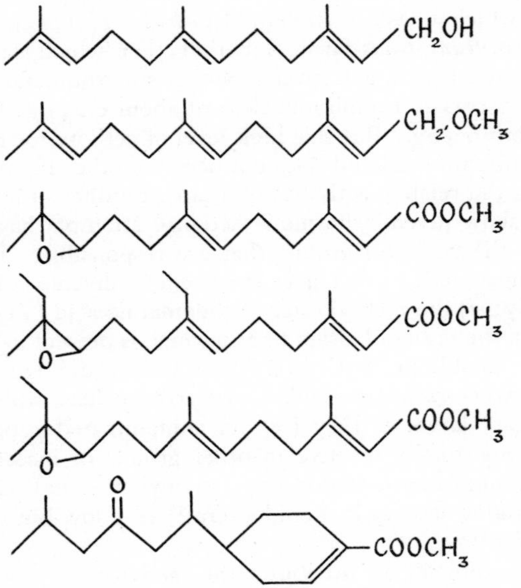

FIG. 23. Formulae of some compounds with juvenile hormone activity. Explanation in the text.

active also in *Rhodnius* but much less so in some other insects, notably in Lepidoptera.

A very interesting compound with widespread activity in many insects, which was synthesized by Bowers et al.[17] was the methyl ester of farnesenic acid with an epoxy ring in the 10–11 position (III). As it turned out, the major component of the active material extracted from *Hyalophora*, which was isolated and identified by Röller et al.[167] proved to be closely similar in structure to compound III, except that it had ethyl groups in place of methyl groups at C_7 and C_{11}; this compound is described as methyl *trans, trans, cis*-10-epoxy-7-ethyl-3, 11-dimethyl-2, 6-tridecadienoate (IV). Along with compound IV there is also another component in the *Hyalophora* extract, isolated by Meyer, Schneiderman, Hanzmann and Ko,[130] which is intermediate between compounds III and IV in having a methyl group at C_7 and an ethyl group at C_{11} (V). This compound is somewhat less active than compound IV.

The major component of the *Hyalophora* juvenile hormone (compound IV) has been synthesized[167] and so have all eight possible geometrical isomers.* The *trans, trans, cis* isomer, which is the natural substance, is by far the most active. This substance is highly active in all the insects on which it has been tested. In *Rhodnius* it has about four times the activity of *trans, trans*-farnesyl methyl ether, which is the next most active substance tested, and gives a perfect supernumerary larval stage (sixth-stage larva) at a minimum dose of about 0·25 μg; farnesyl methyl ether, about 0·9 μg. But this high level of activity for farnesyl methyl ether is quite unusual and does not occur in other insects.[219]

By comparing the relative activities of a large number of synthetic materials which show juvenile hormone activity, attempts have been made to identify the chemical groups that are responsible. But the structure of the most active substances is extremely diverse. To take two examples only: (*i*) the methyl ester of todomatuic acid (VI) which occurs naturally in the resin of balsam fir and which is present in certain samples of paper, notably in the United States, is exceedingly active in the linden bug *Pyrrhocoris apterus* and other Pyrrhocoridae, which produce supernumerary larvae if they rest on contaminated paper; but this substance seems totally inactive in other groups of insects.[17, 183] (*ii*) The straight chain compound dodecanyl methyl ether has relatively high juvenile hormone activity in Lepidoptera;[17] very low but definite activity in *Rhodnius*.[219]

It is therefore difficult to attribute the activity to any single chemical grouping. It was suggested that the juvenile hormone exerts its action by influencing the " permeability relations within the cells—in such a way that the gene-controlled enzyme system responsible for larval characters is brought increasingly into action when the juvenile hormone is present ".[215] It may be that it is the physico-chemical properties of the compounds (relative lipid solubility, polar-apolar balance) as well as the general shape of the molecule as influenced by geometrical isomerism, which determines the reaction with cellular membranes. Such membranes could well show sufficient specific differences to account for the diverse activity of the compounds in different insects.[219] It has indeed been shown experimentally that the substances with the highest juvenile hormone activity are the most effective in depolarizing the cell membrane of the salivary glands in *Galleria*.[2]

* There is an asymmetric carbon atom at C_{11} so that there is a D and an L series of isomers. There is no evidence that the optical isomers differ in activity.

4: Hormones and Reproduction

Role of the corpus allatum in *Rhodnius*

When the corpus allatum becomes inactive the cytoplasm of its constituent cells is reduced, so that the nuclei are separated by vacuolated spaces and the whole gland shrinks (p. 67). This is the state of the gland in fifth-stage larvae of *Rhodnius* at the time of moulting to the adult (Fig. 24A). But it was observed that within a day or so after metamorphosis the corpus allatum enlarges once more and dense cytoplasm again surrounds the nuclei (Fig. 24B). This change is associated with the ripening of a small batch of eggs in the unfed adult female. Undigested blood in the stomach, carried forward from the previous stage, provides the necessary nutrients; when these are exhausted egg formation ceases and the corpus allatum reverts to its shrunken and vacuolated condition. On feeding, it again enlarges and resumes its active appearance (Fig. 24D).[211]

These observations suggested that the corpus allatum was secreting a hormone necessary for the ripening of the eggs. This was confirmed by cutting through the head at such a level that the brain was removed but the corpus allatum remained: unfed newly moulted females, or females that had just taken a meal of blood, matured their eggs normally. But if the corpus allatum was removed by cutting through the neck behind the gland, no development of the oöcytes took place. Parabiosis of such a decapitated insect with another adult which retained its corpus allatum, or implantation of an active corpus allatum into the decapitated female likewise led to complete maturation of the ovaries.

The same visible changes can be observed in the corpus allatum of the adult male; but the secretion has no detectable effect on the formation of spermatozoa. These are matured and accumulated, and distend the vesiculae seminales, whether the insect is decapitated or not. On the other hand the accessory glands, which normally become distended with a clear fluid, remain attenuated and empty in the decapitated male;

whereas they develop normally if the head is transected in front of the corpus allatum or if the decapitated male is joined in parabiosis to insects in which the corpus allatum has been retained.   The hormone appeared to be the same in the two sexes, for it made no difference in the parabiosis or in the transplantation experiments which sex was used

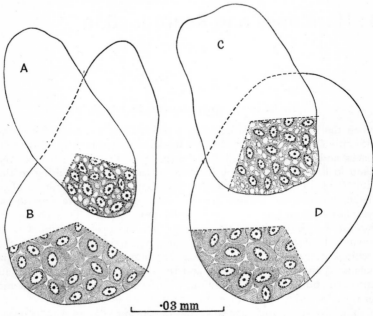

FIG. 24.   Outline and detail of sections of the corpus allatum in adult female *Rhodnius*.   A, immediately after moulting; B, a week after moulting; C, 1 month after moulting (unfed); D, 6 days after feeding.

to induce egg maturation in the female or accessory gland secretion in the male.

In the fasting adult female of *Rhodnius* the oöcytes produced in the germarium at the upper end of the ovariole become clearly recognizable and each continues to grow until the point where it is invested by a layer of follicular cells and yolk is about to be deposited.   Then the nucleus of the oöcyte degenerates and the follicular cells around lose their regular arrangement and begin to invade the oöcyte.   As they move inward they absorb the cytoplasm until finally the interior of the oöcyte is completely filled with follicular cells, many of them breaking down by chromatolysis.   Gradually the entire mass decreases in size and disappears by autolysis; and the next oöcyte to reach the follicular stage suffers the same fate. Precisely the same cycle of partial maturation, death and autolysis takes place in the adult female after feeding and decapitation. On parabiosis with another male or female with the corpus allatum intact, or following the implantation of an active corpus allatum, the oöcyte does not die, but the active follicular cells provide it with nutrients so that the protein and lipid yolk accumulates and the egg-shell is laid down.[211]

Role of the corpus allatum in other insects

Results identical or very similar to those obtained in *Rhodnius* have been reported in many other insects. In the grasshopper *Melanoplus* both deposition of the yolk and egg-shell by the follicular cells, and the secretion in the lower part of the oviduct which provides the oötheca, are dependent on the corpus allatum;[216] and in the cockroach *Leucophaea* both yolk production and secretion in the accessory glands can be restored by re-implantation of the corpus allatum.[212] In *Calliphora*[216] and in *Drosophila*[212] removal of the corpora allata in the adult fly usually prevents the deposition of yolk in the oöcytes. The same applies to the earwig *Anisolabis*,[216] the potato beetle *Leptinotarsa*[216] and the bed-bug *Cimex* after fertilization.[216]

In the mosquito *Culex molestus* the ovaries will develop in the absence of a meal of blood; whereas in *Culex pipiens* and *Aëdes aegypti* a blood meal is always needed. This difference seems to depend upon differences in the secretory activity of the corpora allata, for transplantation of *molestus* corpora allata into *pipiens* hosts will induce egg development in them even in the absence of a blood meal; and transplanted ovaries of *molestus* will develop in *pipiens* or in *A. aegypti* only when these hosts are given a meal of blood.[216]

In *Rhodnius* and in most other insects studied the oöcytes develop normally in the absence of the corpus allatum until they reach the stage when yolk is deposited; but in *Dytiscus*, although the corpus allatum secretion is certainly needed for yolk formation, it seems to be needed also at a much earlier stage of oöcyte development.[218] In some insects the results depend upon the time at which the corpus allatum secretion is supplied. If an active corpus allatum is implanted into the fifth instar of *Locusta* too late to affect metamorphosis, it may accelerate ovarian development. If the corpora allata are extirpated early in the fifth instar the oöcytes in the resulting adult develop up to a length of 0·9 mm and by then contain an appreciable amount of yolk; if extirpated in the adult at 8 days the oöcytes reach a length of 1·4 mm before they cease growing.[216] The same applies also to *Periplaneta*:[216] if the corpora allata are extirpated in the last larval stage, growth of the ovaries is arrested only *after* the beginning of yolk formation; but if extirpated in the ninth or eighth larval stages, arrest occurs *before* yolk formation—because in the precocious adults produced the ovaries are still in a very juvenile condition.

The activity of the corpora allata and the ripening of the oöcytes are both highly dependent on nutrition. That is evident in the observations on *Rhodnius* as described above; it is particularly striking in the milk-weed bug *Oncopeltus* as studied by Johansson. In the adult female of *Oncopeltus*, feeding normally, the corpus allatum increases in volume 12 times during the 2 weeks after moulting; in females given water alone it increases only 2·5 times. If the corpus allatum is removed from a normally fed female and implanted into a female that has been given water alone, she will start egg production despite the lack of food.[216] The same result is seen in *Leucophaea* where again starvation inhibits ovarial development not by a direct effect, but through the corpora allata: starved females will develop eggs if corpora allata from fed donors are implanted.[216] Adult females of the blowfly *Calliphora* ingest preferentially a fluid rich in protein during the early stages of egg growth; whereas during yolk formation protein ingestion declines, carbohydrate feeding increases, and the volume of the corpus allatum is reduced[216]—but here it is difficult to decide which is cause and which is effect.

In the blowfly *Phormia* the clearest effect of the corpus allatum in the adult insect of both sexes is that on lipid metabolism in the fat body. After extirpation of the corpus allatum the fat body hypertrophies.[152] Indeed it is claimed by Orr that " it does not appear necessary to postu-late the existence of an ovarian hormone in *Phormia*. The ovary can simply be regarded as a principal site of utilization of ingested materials."

The corpus allatum in some insects has a striking effect on sexual behaviour. As we shall see in a later chapter (p. 137) in the male *Schistocerca* the corpora allata are involved in inducing production of the pheromone which stimulates sexual maturation in other members of the species[216] and in the male *Locusta* sexual behaviour is totally inhibited after allatectomy.[67] In the normal female of the grasshopper *Gomphocerus*, after the copulatory act there is a sudden change from sexual receptivity to defensive behaviour. This results from the spermatophore which stimulates the receptaculum seminis mechanic-ally. If the receptaculum is denervated or the ventral nerve cord is cut, there is no secondary defence and the female will copulate repeatedly. Allatectomy during the last larval stage or soon after the imaginal moult leads to continuous primary defensive behaviour; and the same happens within 6 days after removal of the corpora allata of mature females.[118] Likewise in *Drosophila*, if mature corpora allata are implanted into pupae just before eclosion the adult female shows a precocious onset of sexual receptivity.[127]

Role of the corpus allatum in Lepidoptera

In *Bombyx mori* the ripening of the eggs is completed in the pupa before emergence of the adult insect and it was shown by Bounhiol[14] that removal of the corpora allata does not prevent the development of eggs. Likewise, if the isolated pupal abdomen of the silk moth *Hyalophora cecropia* is induced to undergo metamorphosis by implantation of prothoracic glands, it will produce eggs.[216]   But even in *B. mori* extirpation of the corpora allata from the fifth-stage larva always results in a decrease of some 12 per cent in the number of eggs produced.[216] In *Galleria* the corpora allata are clearly necessary for egg production. If they are removed at the end of the last larval stage the number of eggs produced is normal; if removed at the beginning of the last stage, the number of eggs falls to 38 per cent of the normal; if removed at the beginning of the penultimate stage, there is a further reduction to 13 per cent of the normal.[166]   These effects can be reversed if other corpora allata are implanted; and similar effects are produced if the corpora allata are inactivated by starvation (sometimes called ' pseudo-allatectomy '). It looks as though in these insects the hormone may exert its action on the ovaries, or the young oöcyte, long before visible development takes place.

These are all species of Lepidoptera in which the eggs are produced in one large batch that is developed during pupal life. It is likely that in other Lepidoptera, in which egg development can depend upon feeding or on water drinking by the adult, the corpora allata may prove to be involved. That is well seen in *Pieris brassicae* in which the ovaries mature after emergence, and in which extirpation of the corpora allata, either in the fifth-stage larva or in the newly emerged adult, arrests the deposition of yolk in the eggs; and in which re-implantation of corpora allata completely restores vitellogenesis.[91]

Less attention has been given to the function of the corpus allatum in the male. In the meal moth *Ephestia* the corpora allata undergo renewed growth in the pupa, and this growth is much greater in the male. During imaginal life there is prolonged secretory activity in the male; in the much smaller corpora allata of the female the secretory process is much less marked.[216] This sexual difference in the size of the corpora allata occurs also in *Hyalophora* adults (in the pupa of either sex the corpus allatum weighs 30 μg; in the adult female about 65 μg; in the adult male about 400 μg). These observations suggest that the hormone has an important function in the male sex.   In the male

Rhodnius, as we have seen (p. 71) it is needed for the activity of the accessory glands, which are responsible for secreting the spermatophore. Adult Lepidoptera produce great numbers of spermatophores in rapid succession.[216] It is probable that that is why the activity of the corpora allata is so conspicuous in male Lepidoptera. The converse is seen in the Coccid *Lecanium* in which the corpora allata of the male are very small objects, whereas in the female they are actually larger than the brain.[10]

Nature of the corpus allatum hormone in the adult insect

When the hormonal function of the corpus allatum in the stimulation of reproduction was first demonstrated in *Rhodnius* it was readily proved that the hormone responsible was distinct from the ' moulting hormone ' and it was assumed also to be distinct from the ' inhibitory hormone ' (= juvenile hormone). But it was pointed out by Pfeiffer[212] that there was nothing in the *Rhodnius* experiments which precluded the identity of these two hormones. She showed that in *Melanoplus* the adult continues to secrete the juvenile hormone and she suggested that the same hormone might be concerned in yolk formation. It was later shown in *Rhodnius* that the ' juvenile hormone ' of the young larva and the ' yolk-forming hormone ' of the adult appear to be interchangeable and probably identical substances.[216] This conclusion is supported by the fact that farnesyl methyl ether and related compounds (p. 68) likewise induce yolk formation as well as maintaining larval characters in *Rhodnius* (though larger quantities of the active principle are needed for yolk formation[216]), and that the corpus allatum of the larva and adult of *Pyrrhocoris* have the same effects on metamorphosis, on metabolism, and on yolk deposition.[182]

In *Schistocerca*, as in *Rhodnius*, etc., the secretion from the corpora allata has a direct effect on the ovary: it seems to ensure the transfer of nutrients by the follicular cells from the haemolymph to the yolk. In the absence of the corpus allatum the oöcytes are resorbed. This resorption can be at least partially prevented by smearing the insects with farnesol.[77] Both cecropia extract and farnesol applied to the surface of the integument of allatectomized females of *Periplaneta* will induce yolk formation.[216]

We saw that the corpus allatum of the adult male *Hyalophora cecropia* is some six times larger than that of the adult female (p. 75) and this is reflected in the amount of juvenile hormone accumulating in the tissues. The adult male contains five times as much ether extractable lipid per gram of wet weight as does the female; and the oil extracted from the female has only one-eighth of the juvenile hormone activity of that from the male. The total content in the male is therefore about forty times that in the female.[64] If ovaries are implanted into the male pupa the extractable juvenile hormone is markedly decreased. It does appear, therefore, that the developing ovaries consume juvenile hormone.[64]

Raison d'être for the hormonal regulation of reproduction

The question arises as to why a hormonal stimulus should be needed by most insects for the full activity of the reproductive system. The reason may well be the same as that suggested for the existence of a moulting hormone (p. 9), that is, because egg production, like moulting, is a cyclical process. It is desirable that it should be initiated only at those times when (*i*) nutrition, (*ii*) the phase of the reproductive cycle, that is, the presence or absence of eggs lower down the tract, and (*iii*) the season of the year, are all appropriate. The insect must therefore have some mechanism for restraining ovary development. A hormonal control system actuated by suitable environmental stimuli and feedback mechanisms is a natural solution.

In the Thysanura where moulting cycles alternate with reproductive cycles throughout adult life, hormonal regulation is clearly necessary.[207] In the parthenogenetic *Carausius* where feeding and egg production are continuous processes in the adult female, there seems to be no call for hormonal regulation, and removal of the corpora allata does not influence yolk formation.[212] But the corpus allatum–juvenile hormone system is not the only system concerned in the regulation of reproductive activity.

Role of the nervous system and neurosecretory cells in ovarial development

The corpus allatum secretion plays an important part in stimulating the deposition of yolk in the oöcytes of *Calliphora*; but occasional females develop their eggs fully even when the corpus allatum has been removed.[212] If the gland is extirpated in the last larval stage of *Calliphora* the resultant females develop their ovaries quite normally and lay viable eggs.[216] In 1952 E. Thomsen discovered that it was possible to see the neurosecretory cells quite readily in the dorsum of the brain in the adult blowfly under dark-ground illumination and to extirpate them by dissection in the living fly. If the medial group of neurosecretory cells was eliminated in this way the oöcytes failed to grow beyond about 0·17 mm. Subsequent implantation of corpora allata induced some slight further growth—but not beyond the initial stage of yolk deposition; whereas re-implanted medial neurosecretory cells led to complete development. The corpus cardiacum serves as a store for the product of the neurosecretory cells and can therefore replace them in the implantation experiments.[212] This discovery has rendered the theory of hormonal control of reproduction in insects considerably more complicated.

Likewise in mosquitoes, in which, as we have seen (p. 73), the corpora allata play an important part in inducing egg development, complete development up to maturity is dependent on a hormone secreted apparently by the neurosecretory cells of the brain.[216] The mosquito *Aëdes taeniorhynchus* is an autogenous species: provided it is adequately fed in the larval stage it produces a batch of eggs in the adult which has not fed after emergence. Both medial neurosecretory cells and corpora allata are essential for yolk formation; but they seem to regulate different processes in egg maturation, for transplanted neurosecretory cells will not restore egg development arrested by allatectomy; nor will implanted corpora allata restore egg development arrested by neurosecretory cell ablation.[107] In *Rhodnius* there is no reduction in the number of eggs produced if the brain is removed, provided the corpus allatum remains.[216] In the stick insects *Carausius* and *Clitumnus*, in which egg production is not affected by removal of the corpora allata, eggs are still produced in the absence of the medial neurosecretory cells, but in smaller numbers.[216] The same is true of *Oncopeltus*. Here the corpus allatum seems to be the essential source of the yolk-forming hormone, but this gland is dependent on connection with the medial neurosecretory cells in the brain for its full activity; and these cells in turn are dependent upon nutrition.[216]

We saw that transplanted corpora allata in *Rhodnius* are fully capable of reproducing the effects of the juvenile hormone both in larval development and in adult reproduction. The gland remains active for a long time after removal—but not indefinitely (p. 50). In the mirid bug *Adelphocoris*, removal of the corpus allatum arrests the growth of oöcytes, but implantation of corpora allata into the operated females will not restore egg maturation because the implanted glands quickly lose their secretory activity; continuous nervous connection with the brain seems necessary.[52] Likewise in *Tenebrio*, denervated corpora allata do not secrete hormone. The corpus allatum normally facilitates yolk deposition; but the neurosecretory cells are needed (*a*) for the production of mid-gut proteases and probably other proteins, and (*b*) for the activation of the corpora allata by axon pathways.[135]

Peripheral nerves also may influence ovarian development. In the mosquitoes *Culex pipiens* and *Aëdes aegypti*, prolonged distension of the gut and the abdomen by sealing the anus will lead to egg development even in the absence of a blood meal;[216] and the production of proteolytic enzymes in the gut of the tsetse fly *Glossina* that occurs during the twenty-four hours after eclosion appears to depend on the inflation of the crop with air.   This may well be the result of neuroendocrine control regulated through the stomatogastric nervous system (p. 90).[103] In the bed-bug *Cimex* the corpus allatum does not become active, even in the fed female, until the sperm, after passage through the body cavity, have reached the correct location in the female genital tract; and soon after the sperm have disappeared from the spermathecae, the ovary reverts to the virgin condition.[216] This stimulus to the corpus allatum does not appear to be of a hormonal nature; for although removal of the receptaculum seminis of the mated female results in inactivation of the corpus allatum, the implantation of the receptaculum from a mated female into a virgin female does not cause activation.   On the other hand, cutting the nerve cord within 3 hours after mating does prevent activation of the corpus allatum.[45]

We are concerned with a complex system comprising neurosecretory cells in the protocerebrum and suboesophageal ganglion, the corpus allatum, the corpus cardiacum, and the ovary.   These components influence one another by nervous stimuli, by neurosecretions carried along nerve pathways, and by humoral factors in the circulating blood. The relative parts played by the different elements in this system clearly vary from one insect to another.   They have been most closely studied in cockroaches and in grasshoppers.

Regulation of reproduction in cockroaches

In *Leucophaea maderae* the brain is the final regulator of the function of the corpus allatum throughout the reproductive phase.   The species is ovoviviparous (the eggs are retained in the brood chamber until the moment of hatching) and the ovaries become inactive during pregnancy. This inactivity results from the arrest of secretion in the corpus allatum. If active corpora allata from larvae are implanted into pregnant females the ovaries resume their development; the oöcytes ripen and yolk is deposited.   The corpus allatum (and thus in turn the ovary) can be activated also during pregnancy by cutting the corpus allatum nerve or the medial corpus cardiacum nerve, or by destroying a certain part of the protocerebrum.   This is effective even when the neurosecretory cells and the pathways for the neurosecretion are untouched.

It appears that during pregnancy a nerve centre in the brain inhibits the corpus allatum; but the activation of the corpus allatum by these operations takes place only when the nervous connection between the suboesophageal ganglion and the corpus allatum remains intact.   The corpus allatum cannot continue to function if all nervous connections are severed.[50]   It will be recalled that a similar inhibition of the corpus allatum comes into operation during the metamorphosis of the last larval stage (p. 50).

More recent experiments have given results which differ somewhat from those outlined above; for it has been found that in pregnant females of *Leucophaea* not only are the corpora allata inactive but during the first twenty days of the pregnancy period the ovaries appear to be incompetent for oöcyte development.   Implantation of active corpora allata has no effect on them and even if the ovaries are transplanted to females with developing ovaries they fail to grow.[175]

The mechanism by which the presence of developing eggs in the lower genital tract brings about nervous inhibition of the corpus allatum by the brain has been a subject of controversy.   Evidence obtained in the cockroaches *Blattella* and *Pycnoscelus*[216] and *Diploptera*[216] as well as in *Leucophaea*[50] has shown that the presence of an oötheca in the brood chamber provides a mechanical stimulus which can be simulated by the insertion of a dummy oötheca made of glass.   This stimulus will be conveyed to the brain by the ventral nerve cord.   If the oötheca is removed, or the nerve cord of the pregnant female is cut (with the oötheca still in position) inhibition of the corpora allata ceases and they become activated.

A further stimulus that can intervene in the hormonal control of reproduction is the act of mating.   In the viviparous cockroach *Diploptera* mating seems to be essential for the normal rate of egg growth; and since allatectomy after mating prevents egg maturation the stimulus connected with mating presumably activates the corpus allatum.[50]   The stimulus to the corpus allatum is probably nervous, for section of the abdominal nerve cord prior to mating also stops egg growth.   On the other hand, if the corpora allata are separated from the brain in virgin females they become prematurely active and several successive batches of eggs may be developed.

It appears that the corpora allata are normally restrained by the brain and that mating stimuli counteract this restraining influence. During pregnancy the corpora allata are similarly restrained, and they resume their activity only in response to renewed afferent nervous stimuli, normally provided by the act of parturition.   Egg maturation can also be induced by inserting an artificial spermatophore in the form of a glass bead into the bursa copulatrix or by excising the gonopophyses of virgin females.[50]   Thus mating (in virgin females) or parturition (in the pregnant female) acting by way of the ventral nerve cord, brain and corpus allatum can provide an adequate stimulus to induce ripening of the eggs.[50]

In decapitated females of the cockroach *Nauphoeta cinerea* oöcyte growth is arrested at once while the respiratory rate of the fat body remains high for two days and then drops to half the original value and protein synthesis to one-third that of the controls. If corpora allata are implanted in these decapitated females normal oöcyte growth is induced and respiration and protein synthesis in the fat body remain at the normal level. These metabolic effects of the corpus allatum are observed also in females that have been castrated as well as decapitated. Implanted brains and corpora cardiaca were not effective, save for a very slight stimulation of respiration. It has therefore been concluded that in this cockroach the corpus allatum is responsible both for stimulating the synthesis and release into the haemolymph of proteins and other nutrients, and for the transport of these nutrients through the follicular epithelium to the oöcytes.[122]

Regulation of reproduction in locusts

The hormonal control of reproductive physiology has been analysed in considerable detail in locusts. In *Locusta* each batch of eggs represents the terminal oöcytes. These batches succeed one another at intervals of 6–7 days; but the interval between successive oöcytes is only $2\frac{1}{2}$ days. Clearly the presence of ripe eggs inhibits the maturation of subsequent oöcytes.[216] As Highnam points out, the linkage of the brain, by way of the corpus allatum, with ovarial development, is common to many insects, but the part played by the brain in this process may be either humoral or nervous.[216] In *Schistocerca*, cauterization of the neurosecretory cells, or removal of the corpora cardiaca, prevents growth of the terminal oöcytes; re-implantation of neurosecretory cells into such females gives rise to normal development.[216] The secretion from the brain is evidently important for ovarial growth.

In *Locusta*, removal of the frontal ganglion will prevent sexual maturation in both sexes; but there is some doubt as to whether this is a simple effect of lack of nutrients: extirpation of the frontal ganglion interferes with feeding and swallowing; or whether the frontal ganglion serves as a relay station between the intake of food and the renewed synthesis of neurosecretory material in the brain.[36, 79] Clarke and Gillot[35] provide good support for the view that removal of the frontal ganglion does not merely cause starvation but results in defective liberation of the neurosecretory substance from the corpus cardiacum and thus leads to defective protein synthesis, particularly the synthesis and release of proteolytic enzymes. This defect can be made good and normal growth restored by the injection of corpus cardiacum extracts. Removal of the frontal ganglion has much the same effect as cauterizing the neurosecretory cells: the corpora allata fail to enlarge and the oöcytes fail to mature.[195]   The general conclusion seems to be that removal of the frontal ganglion commonly inhibits egg development—but this is a nervous effect via the neurosecretory cells and not a direct endocrine effect.[171]

In the cockroach *Blaberus* electrical stimulation of the brain causes depletion of the neurosecretory material in the protocerebrum.[216]   The same change occurs in *Schistocerca* if the optic nerves are stimulated with high frequency electric shocks; indeed a wide variety of stimuli, such as rotating the female in a flask, drastic wounding, or copulation, all have the effect of emptying the neurosecretory cells and corpora cardiaca of stainable material—and all these treatments lead to accelerated development of the oöcytes.[216]   In *Schistocerca* copulation is necessary if the full complement of eggs is to be laid; and repeated copulation is needed for successive batches of eggs—although a single copulation will supply all the spermatozoa that are required.[216]   It seems likely that copulation provides the normal stimulus for the liberation of the neurosecretory material.

All the evidence suggests that neurosecretory cells which are laden with stainable secretion are in the inactive state; partially emptied cells are active.[216]   The fact that DL-cystine labelled with ^{35}S is taken up by the neurosecretory cells much more strikingly in females reared with mature males, supports the conclusion that when the system contains little stainable material the hormone is being released most rapidly. For the neurosecretory system contains little stainable material in females whose terminal oöcytes are developing rapidly in the presence of mature males, but contains a large amount when the terminal oöcytes are mature, or in females reared without males, whose terminal oöcytes are developing very slowly.[216]

In *Locusta* and *Schistocerca* the neurosecretory cells in the central region of the pars intercerebralis, whose product is liberated via the corpora cardiaca, seem to be concerned in controlling protease activity in the mid-gut and the protein content of the haemolymph; the failure of oöcyte development after removal of this region can be traced back to the low availability of protein.[78, 117]   The neurosecretory cells which occupy the lateral regions of the pars intercerebralis supply the corpora allata; their extirpation inhibits corpus allatum activity (Fig. 26); and since the corpus allatum secretion is essential for oöcyte growth this also is arrested.[67, 90, 194]

It seems that the neurosecretory cells, the corpora cardiaca and the corpora allata act as a unit in the control of reproduction and the relative importance of the components in this unit seems to vary in different insects.   In *Calliphora* the neurosecretory system may influence ovarian development both directly and by the production of a hormone acting on the corpus allatum.[216]   In Dictyoptera and Orthoptera, as we have seen, the neurosecretory system provides a link between the nervous and endocrine moieties of the coordination mechanism.[216]   In *Rhodnius*[211] and *Oncopeltus*[216] the juvenile hormone seems essential for the survival and ripening of the oöcytes and appears to be the dominant factor in the endocrine control of reproduction. But even here the neurosecretory cells are probably playing a part.[44] Further consideration will be given to this problem in discussing the metabolic effects of hormones (p. 105).

Control of ovulation and oviposition

Humoral factors, probably derived from the neurosecretory cells in the brain, are involved also in ovulation and oviposition.   The blood of fecundated *Bombyx mori* will induce oviposition when injected into unfertilized females,[216] and it is at the time of oviposition that the neurosecretory material in the adult female *Bombyx* is most actively discharged along the axons.[216]   In *Dermestes*, mating acts as a stimulus to oviposition and appears to hasten final maturation of the eggs;[216] and in *Aëdes aegypti* females, also, ovulation seems to be provoked by some humoral factor which is liberated when the spermathecae are filled with sperm.[216]

In the Pyrrhocorid bug *Iphita limbata*, as studied by Nayar, the neurosecretory cells of the pars intercerebralis seem to be concerned in oviposition.   Immediately before oviposition no visible neurosecretory material can be seen entering the corpus allatum from which the supply seems to be cut off; but instead it flows into the anterior end of the aorta close to the corpus cardiacum.   Meanwhile the material in the neurosecretory cells becomes depleted; and in females which have started oviposition the cells are almost devoid of stainable material.   Blood from a laying female will induce oviposition when injected into a female not yet fully gravid; but blood from earlier stages, or from a female which has started mating again, has not this effect.[216]

It may be that an ' ovarian hormone ' is concerned in this response, for injection of a watery extract of the ripe ovary will likewise stimulate the neurosecretory cells to discharge their secretion and so induces immediate oviposition.   The role of the neurosecretory cells in oviposition in *Iphita* has been confirmed by implanting clusters of these cells into nearly gravid females: they likewise induced immediate oviposition.[216]

In the published work on this subject it is not always clear, when an ' increased fecundity ' or an ' increased egg production ' are described, whether there is an increase in the maturation of eggs or whether it is the ovulation and oviposition of ripe eggs which is being stimulated. In the suboesophageal ganglion there are often neurosecretory cells which become conspicuously enlarged in the female after castration. In *Gryllus* and *Carausius*, extracts of the suboesophageal ganglion stimulate the deposition of already formed eggs.[199]   In *Rhodnius* the rate of egg laying is greatly increased by mating, but this has little effect on the rate of egg formation.[38]   The stimulating action of mating occurs only if sperm are introduced into the spermathecae: it does not happen if the opaque accessory glands of the male are removed so that the spermatozoa fail to reach the receptacula (p. 103).   Since implanted spermathecae from normally mated females will induce increased egg laying by virgin females, it was concluded that a blood-borne humoral factor from the spermathecae provides the stimulus.[44]   It probably acts via the brain, since surgical removal of the neurosecretory cells abolishes the effect.[44]   But in the cockroach *Pycnoscelus* the sperm exerts this influence on oviposition not by humoral means but by way of the nervous system: innervated spermathecae are required for successful oviposition.[187]

In *Drosophila* the increased egg laying that follows mating seems to depend on a secretion from the accessory glands (paragonia) for implantation of the vesiculae seminales from males into virgin females had no effect, whereas implantation of paragonia did.[61]   This is certainly an effect on ovulation and deposition of eggs retained in the ovarioles of virgin females; and the implanted paragonia also have the effect of depressing the sexual receptivity of the female.[129]   Similar results are reported for the mosquito *Aëdes aegypti*; indeed the active substance from the paragonia of the male mosquito can be replaced by glands from *Drosophila*, but not by those from *Tenebrio*. The extract is inactivated by heat but not by freezing.[108]

Hormones and diapause in the reproductive stage

Reproduction, like growth, may suffer a periodic arrest; and as with growth (p. 28) this arrest may be the direct effect of an adverse environment, or it may be a true ' diapause ' which persists even under favourable conditions.   In *Dytiscus* and various other beetles, the gonads revert to a resting state after the first reproductive period; they show renewed activity about the same date the following year, sometimes in a third or fourth year also.   This seems to be a deep-seated rhythm and not a simple effect of warmth following the winter cold; for in *Carabus coriaceus* and *Leistus* spp. the breeding season does not begin until the late summer or autumn,[218] and *Dytiscus marginalis* lays its eggs in March, *D. semisulcatus* in October.[218]

These reproductive cycles are probably controlled by hormones. The full development of the eggs in *Dytiscus marginalis* is dependent on the corpus allatum.[218] Reabsorption of the oöcytes takes place in July as soon as they have grown to 0·8 mm in length; in February they reach 2 mm before they break down; and in March they attain the normal length of 7 mm. But if corpora allata are implanted, the full development of the eggs can be produced at any season; and if the corpora allata are removed at the height of egg production, the ovaries regress. In Carabids, as in *Dytiscus*, the corpus allatum hormone prevents degeneration of the oöcytes and permits ripening of the ovaries.[218]

The hormonal regulation of adult diapause has been studied in great detail in the Colorado potato beetle *Leptinotarsa*. The induction of diapause is a short-day effect which supervenes in the late summer and autumn. It is associated with clear-cut changes in behaviour and metabolism: the beetles burrow into the soil where they come to rest; the rate of oxygen consumption falls to a low level; and ovarian development is arrested. These changes can be reproduced by surgical removal of the corpora allata, and there is little doubt that their immediate cause is the suspension of juvenile hormone secretion.

Under long-day treatment the corpora allata are activated, probably by hormonal stimuli from the medial neurosecretory cells. Short-day treatment, on the other hand, not only inactivates the neurosecretory cells and so removes the hormonal stimulation of the corpus allatum but there is also active inhibition of the corpus allatum by way of their nerve connections.[221, 222] Diapause in *Leptinotarsa* can also be induced by feeding an aged potato foliage. This is not a direct effect of poor nutrition but is again due to neuroendocrine arrest brought on by sensory information about the host plant. The corpora allata are inhibited; but reproduction can be restored by the implantation of active glands even though the food remains unchanged.[223]

In hibernating adults of *Leptinotarsa* the flight muscles show extreme degeneration: the muscle fibres are reduced in diameter and the mitochondria, or sarcosomes, are virtually absent. Similar signs of degeneration follow extirpation of the corpora cardiaca and corpora allata. If several of the active complexes are re-implanted there is a very rapid regeneration of muscle fibres and a new formation of mitochondria as the beetles come out of diapause.[189] Under normal conditions the beetle works its way into the superficial layer of the soil where it rests until the capacity for flight is restored. Quantitative assays for juvenile hormone in the blood of adult *Leptinotarsa* show that this falls to a very low level during diapause, but rises again during long-day treatment.[222]

Very similar results have been obtained on the Chrysomelid beetle *Galeruca tanaceti*; here the use of ^{35}S-DL-cystine in autoradiographic studies has confirmed that the rate of turnover of cystine in the neurosecretory cells increases about two-fold when a mature female begins to oviposit and about ten-fold when a female goes from diapause to oviposition. In diapausing females the labelled cystine builds up in the neurosecretory cells; in ovipositing females it accumulates in the corpus cardiacum and is discharged within 24 hours.[180]

Mosquitoes become more or less immobile in the autumn; and after taking meals of blood they tend to develop an enlarged fat body instead of maturing their ovaries, a condition known as ' gonotrophic dissociation '. In *Culex pipiens*, this change seems to be closely related with the temperature: if given blood and kept warm at any time in the winter they will develop eggs. But if *Anopheles maculipennis* race *atroparvus* is fed with blood during the winter and kept at 29°C, although some develop eggs, others lay down reserves in the fat body; and *A. maculipennis* race *messeae* in the autumn shows complete ' gonotrophic dissociation ' and fails to produce eggs at any temperature.[218] In this state there is no absorption of developing oöcytes as described above (p. 72); the ovaries show no signs of growth at all.

The cessation of blood-feeding in *Culex pipiens*, and the gonotrophic dissociation in *Anopheles maculipennis* during the winter diapause are the result of exposure to a short (12-hour) photoperiod.[42] The changes are doubtless associated with arrest of secretion in the neurosecretory cells and in the corpus allatum—but the details have not been described. The same is probably true of the reproductive arrest that occurs in many other insects, such as *Polistes*, where the neurosecretory cells of the pars intercerebralis become laden with neurosecretory material during the winter diapause and emptied again when diapause comes to an end,[193] *Musca autumnalis*,[192] the locusts *Nomadacris*[195] and *Anacridium* where again the neurosecretory cells are filled with secretion during diapause and the corpora allata are small and inactive,[62] and in the alfalfa weevil *Hypera* where topical application of the methyl ester of 10, 11-epoxyfarnesenic acid (p. 69) effectively terminates the summer diapause.[16]

An unusual example of an environmental factor affecting hormone secretion is seen in the rabbit flea *Spilopsyllus cuniculi*. The ovaries of fleas kept on male or non-pregnant female rabbits remain immature as in diapause; whereas those on a pregnant host are mature by the time the young are born. The factor involved is available only during the final week of pregnancy and not at all in male or non-pregnant female rabbits. The factor disappears from the female after parturition but is present in her nestlings for at least 7 days. The factor does not have solely a trigger-like action in initiating ovary development; the ovaries undergo regression in maturing fleas transferred to hosts that do not supply the factor. It is suggested that the ' yolk-forming hormone ' normally secreted by the corpora allata may only be produced by rabbit fleas when they obtain the postulated factor. Whether the steroid hormone levels in the host are involved is still uncertain.[128, 170]

5: Neurosecretion and Neurohumours

It has long been recognized that the neurones of insects, like those of other animals, are in a state of more or less continuous secretory activity. The secretory product is the axoplasm which, along with the mitochondria that it contains, seems always to be in slow motion away from the cell body: at points where a wide axon becomes constricted, as in the ganglia of *Rhodnius*, the mitochondria may be dammed up. But in this chapter we shall be concerned with nerve cells as a source of hormones.

Some forty years ago it was observed that certain nerve cell bodies, in both vertebrate and invertebrate animals, showed histological staining properties which suggested that they were concerned in the production of special secretions. The leading exponent of this idea was the late Ernst Scharrer who may be said to have founded the theory of 'neurosecretion'. This interpretation is now universally accepted.[60]

But neurones exercise another chemical function. They liberate pharmacologically active substances at the axon endings. Such substances, which are commonly called ' neurohumours ', are believed to play a part in the transmission of impulses across synaptic junctions between one nerve process and another and at the endings of axon branches on the muscle fibres. Some neurohumours, such as noradrenaline in vertebrates, may be set free into the circulating blood and produce systemic effects elsewhere in the body.

Neurosecretory cells

The recognition of neurosecretory cells by histological means preceded any experimental demonstration of their physiological action. But it should be recalled that Kopeč had suggested that the brain was the source of the hormone that caused pupation in *Lymantria* (p. 1); and we have already seen (p. 3) that soon after the presence of neurosecretory cells had been discovered in the dorsum of the brain in *Rhodnius* they were identified as the source of the ' moulting hormone ' —later the ' brain hormone ' or the ' activation hormone ' which evokes secretion of ecdysone in the prothoracic gland.

The most characteristic example of these neurosecretory cells elaborates an acidophil product, staining with acid fuchsin or with phloxin, which becomes basophil after permanganate oxidation and then stains a deep blue-black with chromehaematoxylin (Plate IIIA). The product also stains strongly with the paraldehyde-fuchsin of Gomori— usually to give a purple colour (Type A) sometimes a green colour (Type B).[60] On the basis of their staining properties neurosecretory cells have been grouped into four classes; but there are examples of evident neurosecretory cells which cannot be fitted into this scheme; indeed, in a study of the neurosecretory cells in the cellar beetle *Blaps*, Fletcher[55] has shown the presence of no less than thirteen types. Moreover, ganglion cells are frequently described whose staining properties are equivocal; it is uncertain whether they are neurosecretory cells or not. Other useful stains are alcian blue or Victoria blue after performic acid oxidation, which can be used effectively to show up the Type A neurosecretory cells in whole mounts of the ganglia.[47] In their general morphology the neurosecretory cells are essentially identical with the ordinary neurones.

As was first shown by E. Thomsen[216] the neurosecretory cells, and the axons coming from them, when examined in the fresh state under dark-ground illumination, have a luminous blue appearance. This was clearly a ' physical colour ' (Tyndall blue) and indicated that the material was made up of fairly uniform minute particles capable of scattering the shorter wave-lengths of light. This was confirmed by the examination of sections under the electron microscope (originally by Bargmann in vertebrates) which showed that the neurosecretory material is always in the form of minute spheres ranging from about 100–300 mμ in diameter[216] (Plate IIIB). Most of these granules, both large and small, are electron-opaque; but among them are often electron-lucent vesicles of about the same size which closely resemble the ' synaptic vesicles ' that are abundant in the terminations of ordinary nerves. Both types of granule, when massed in the neurosecretory cells of *Bombyx mori*, stain with chrome-haematoxylin and paraldehyde-fuchsin.[1]

These granules appear to consist mainly of protein without associated lipid. As with many protein secretions they are budded off from the Golgi complex and are channelled into the axons and carried along them.[140] The neurosecretory cells of *Calliphora* have a well developed granular endoplasmic reticulum which is doubtless concerned in the synthesis of the protein.[11] This protein, at least that coming from the

Type A cells, is rich in cystine which is responsible for many of the staining reactions, notably the performic acid–alcian blue method of Adams and Sloper.[177]   The product of the Type B cells (in *Dysdercus*) is practically devoid of cystine: it fails to stain in performic acid–Victoria blue.[47]   The relation between this neurosecretory product that is histologically detectable and the active principle of which it is the carrier resembles the relation between the zymogen granules of the pancreatic cells and the enzyme activities that are concomitantly elaborated.[60]   The multiple active substances, presumably of neurosecretory origin, that are liberated from the corpus cardiacum of *Periplaneta* (p. 103) seem mostly to be small peptides, heat stable and dialysable but destroyed by chymotrypsin.[19, 138]

Distribution of neurosecretory cells

Neurosecretory cells are widely scattered throughout the central nervous system of insects.   The medial and lateral groups in the dorsum of the protocerebrum, which have been considered in relation with moulting (p. 5) and reproduction (p. 77), seem to be present in all insects.   Their numbers range from a dozen or so in Hemiptera, at least forty in *Hyalophora*,[75] up to about 2,400 in *Schistocerca*.[216] There are also small groups in the deutocerebrum and tritocerebrum. There is often a pair or more in the suboesophageal ganglion.   Others occur in the ganglia of the thorax and throughout the ganglia of the abdominal chain.

The neurosecretory cells are intimately associated with the visceral or sympathetic nervous system.   The ' stomatogastric ' part of this system, which arises in the course of development from the stomodaeum and only later becomes connected with the brain, is closely concerned in the process of neurosecretion (Fig. 25).   It consists of a median ' frontal ganglion ', joined by bilateral connectives to the front of the brain, which sends backwards a median ' recurrent nerve ' that passes between the brain and the oesophagus to end in a paired or single ' ventricular ganglion ' towards the hind end of the oesophagus; and on its course there is sometimes also a median ' hypocerebral ganglion ' just behind the brain.   But the most important component in relation to the endocrine system is the ' corpus cardiacum '.   This is a paired or fused median structure behind the brain, likewise nervous in origin, which receives by way of two nerves (corpus cardiacum nerves I and II) the neurosecretory axons from the medial and lateral groups of neurosecretory cells in the protocerebrum.   A third nerve (c.c. nerve III) brings neurosecretory axons from the cells in the tritocerebrum.

The histological structure of the corpus cardiacum was for long completely baffling, but the electron microscope has revealed that it consists of three main components: (*i*) the bulbous ends of the neurosecretory axons from the brain, filled with neurosecretion in the form of the granules described above (Plate IIIc); (*ii*) intrinsic neurosecretory cells the axons of which run out into the nerves of the stomatogastric system to supply the gut and the dorsal aorta; (*iii*) glial cells

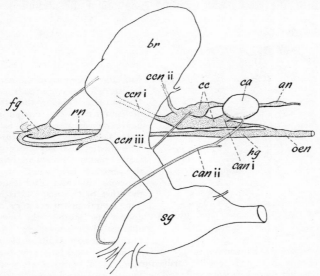

FIG. 25. The brain and suboesophageal ganglion, the stomatogastric nervous system (shaded) and the corpus allatum in the cockroach (after Willey, simplified). *an*, aorta nerve; *br*, brain; *ca*, corpus allatum; *can* i and ii, the two corpus allatum nerves; *ccn* i, ii and iii, the three corpus cardiacum nerves; *fg*, frontal ganglion; *hg*, hypocerebral ganglion; *oen*, oesophageal nerve; *rn.* recurrent nerve; *sg*, suboesophageal ganglion.

which provide the sheaths wrapped around these various components.[216] The corpus cardiacum is also connected by nerves to the corpus allatum, which thus receives some of the neurosecretory axons from the brain; indeed in *Schistocerca* the corpus allatum receives an independent supply of neurosecretory axons from the lateral neurosecretory cells of the protocerebrum[194] (Fig. 26). In some insects there are nerves connecting the corpora allata with the suboesophageal ganglion.

The second part of the visceral nervous system consists of the median or 'unpaired ventral nerves' which arise from the suboesophageal and segmental ganglia and give rise to transverse nerves supplying the spiracles of the same segment. It was noted by Newport (1834) in his classic work on the privet hawk moth *Sphinx ligustri* that the transverse nerves show fusiform swellings which he compared with the sympathetic ganglia of vertebrates. These thickenings have been shown by Raabe[160] to have a structure resembling that of the corpus cardiacum and to be laden with neurosecretory deposits derived from the segmental ganglia. They have been called ' perisympathetic organs '.

Lastly, the compound terminal ganglion of the abdomen, which

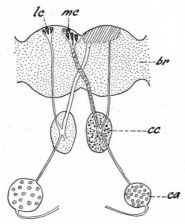

FIG. 26. Diagram of the neurosecretory axons from the brain of *Schistocerca* showing the effects of cauterizing the medial and lateral neurosecretory cells on the right hand side (after Strong). The medial neurosecretory cells supply neurosecretory material to the corpus cardiacum of the opposite side and after cautery the store of this material is lost. The lateral neurosecretory cells appear to supply the corpus allatum of the same side. After cautery the corpus allatum on this side becomes shrunken but on the opposite side remains large. *br*, brain; *ca*, corpus allatum; *cc*, corpus cardiacum; *lc*, lateral neurosecretory cells; *mc*, medial neurosecretory cells. The shaded area has been cauterized.

innervates the sexual organs and the posterior segments of the gut, is commonly regarded as giving rise to the third part of the visceral nervous system. Many of the axons in this system likewise carry neurosecretory granules.

Site of discharge of the neurosecretory product

In insects, as in vertebrates, the product of the neurosecretory cells is often conducted to a specific organ, such as the neurohypophysis of vertebrates, which serves as a gateway for the escape of the active principle from the neurones to the circulating blood. Structures of this kind are called ' neurohaemal organs '. The corpus cardiacum is a classic example. The neurosecretory granules are carried along the axons to the bulbous endings which come to lie just below the surface membranes of the corpus cardiacum, where they are extruded by exocytosis, or ' reverse pinocytosis '. It may be that the granules can only be discharged when a nerve impulse depolarizes the axon membrane and makes it possible for the granule membrane to fuse with the cell membrane.[140] In *Carausius* this appears to happen in regions where a glial sheath is absent.[185]

This interpretation is supported by the observations that in *Blaberus*[216] and in *Periplaneta*[68] neurosecretory cells and axons do transmit nerve impulses, that electrical stimulation leads to the discharge of neurosecretion from the cells, and that high concentrations of potassium, which might be expected to simulate the depolarization by nerve impulses, likewise leads to depletion of neurosecretory material.

We saw that in the Pyrrhocorid bug *Iphita* (p. 83) the neurosecretory axons from the pars intercerebralis end in the wall of the aorta where their secretion is liberated into the haemolymph. A similar arrangement occurs in *Ranatra*. Here there are two medial groups of neurosecretory cells, each with 9–10 neurones, and two lateral groups, each with 3–4 neurones. The neurosecretory axons are separated into two groups: those bearing A type secretion by-pass the corpus cardiacum and end in the wall of the aorta, while those bearing B type secretion end as usual in the corpus cardiacum.[47]

The ' perisympathetic organs ' which form the swellings on the branches of the unpaired ventral nerves seem to be the neurohaemal organs for the neurosecretory cells of the segmental ganglia. Here also the terminations of the axons have no glial covering.[160]

But neurosecretion does not take place exclusively by way of specialized neurohaemal organs. In the suboesophageal ganglion of *Carausius* there are two ventral neurosecretory cells which give off axons that ramify in the dorsal region of the ganglion. Here the secretion collects and is discharged into the haemolymph so that the ganglion itself serves as a neurohaemal organ.[155] Many neurosecretory axons run directly to the organs themselves. Some of the axons from the pars intercerebralis end among the cells of the corpus allatum; for example, in the last stage larva of *Leptinotarsa*.[216] In Blattidae and Culicidae all parts of the vegetative nervous system contain secretion.[216] In Aphids the axons from the intrinsic neurosecretory cells of the corpus cardiacum run into the sympathetic nerves supplying the gut and the heart, and end freely on the muscles of this organ in contacts resembling synapses;[15] and in the lateral cardiac nerves of *Periplaneta* there are neurosecretory axons from the segmental nerves of the ventral ganglia and also neurosecretory cells which likewise give rise to neurosecretory axons—all of which end in neuromuscular junctions on the heart alongside the endings of the motor nerves.[89, 131, 132] Neurosecretory axons from the terminal ganglion of the abdomen, supplying the hindgut and the rectal glands, likewise terminate in the proximity of the organs supplied.[69a] In the case of the diuretic hormone in *Rhodnius* (p. 99) the axons from the neurosecretory cells located in the fused thoracic and abdominal ganglia enter the nerves which fan outwards as they run to all parts of the abdomen. On the course of these nerves the neurosecretory axons give off branches which end in bulbous enlargements immediately below the nerve sheath (Fig. 27). These are clearly the sites of discharge.[124]

All these varied observations can be summarized by saying that the

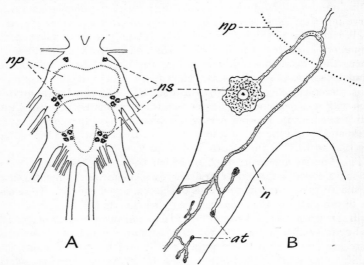

FIG. 27. A, diagram of the fused meso- and metathoracic and abdominal ganglia of *Rhodnius* showing the three paired groups of neurosecretory cells. B, diagram of a neurosecretory cell of the posterior group showing the course of the axon and the swollen endings filled with neurosecretory material which it gives off (after Maddrell). *at*, axon terminations; *n*, abdominal nerve; *np*, neuropile; *ns*, neurosecretory cells.

gateways for the escape of the neurosecretory material may be located in definite neurohaemal organs or at individual nerve endings.[53] Discharge is not restricted to the fibre terminations. Sites of release are characterized by clusters of small electron-lucent vesicles near the internal surface of the plasma membrane, which bear a superficial resemblance to the presynaptic vesicles in interneuronal junctions.[174]

It is generally assumed that the neurosecretory cells are modified motor neurons whose secretion passes along the axon. But in the stick insect *Carausius* and in the blowfly *Phormia* there are a number of multipolar neurones located peripherally in the abdomen which resemble the neurones associated with stretch receptors. They are probably sensory neurones in origin, but they contain typical neuro-secretory granules and these are almost certainly released into the blood. They are transmitted, and presumably released, not by a single axon process, but by multiple, peripheral dendrites.[53] It remains to be discovered whether other parts of the sensory nervous system have a neurosecretory function.

Functions of neurosecretory cells

The physiological functions of the neurosecretory cells for the most part still await discovery. Even the activities of the cells of the pars intercerebralis concerned in moulting and in reproduction, which have been studied most intensively, are still incompletely known (p. 106). The function of many groups of neurosecretory cells is entirely hypothetical.

Metabolism

The products of neurosecretory cells commonly bring about changes in metabolism. Examples of such effects will be discussed in Chapter 6, p. 104, where we shall consider to what extent such changes are primary effects of the hormones or secondary consequences of their primary action. Such considerations are particularly relevant to the neurosecretory influence on moulting and reproduction.

Activity rhythms

The physiological activities of many living cells is subject to an inborn diurnal rhythm—often called a ' circadian ' rhythm because its period is usually a little more or a little less than twenty-four hours in any given individual. These rhythms, whether they concern colour change, oviposition, enzyme secretion, locomotor activity, etc., are under the control of the environment acting through sense organs responding usually to cycles of light or temperature. One of the most closely studied examples is the locomotor rhythm in the cockroaches *Periplaneta* and *Leucophaea* which normally become active as night approaches but whose activity rhythm will continue for some days in continuous light or in continuous darkness.

Janet Harker[218] produced evidence that the change over from rest to activity in *Periplaneta* was the result of an hormonal stimulus in the circulating blood and that the source of the hormone concerned is the group of neurosecretory cells in the suboesophageal ganglion. Some of the experimental results on which these conclusions were based could not be confirmed by Roberts; and in more recent work Nishiitsutsuji-Uwo and Pittendrigh[139] suggest that the neurosecretory centre concerned is located in the pars intercerebralis of the protocerebrum. The precise details of this control system have not been worked out but the prime environmental stimulus seems clearly to be the light cycle experienced by the eyes. It may be, as suggested by Brady,[18] that the various reported results could be accounted for if the cockroach has an electrical pace-maker in the brain coordinating rather ephemeral neuroendocrine rhythms in the nerve cord ganglia. In the cricket *Acheta* the rhythm of activity is lost under constant conditions of illumination but is re-established in alternating conditions if either the compound eyes or the ocelli are covered, but not if both are covered.[142]

Colour change

Two types of colour change are recognized in insects: there may be differences in the relative amounts of different pigments deposited in the epidermal cells or, in the case of melanin, laid down in the substance of the cuticle. The persistent change that results is commonly called a ' morphological colour change '. Alternatively, in such insects as Mantids and Phasmids the pigment granules present in the epidermal cells may change their distribution and this results in a transient or ' physiological colour change ' which is readily reversible.

Morphological changes in pigmentation occur in many insects in response to the environment. Pupae of butterflies (*Pieris, Vanessa*, etc.) and certain caterpillars can vary their coloration according to the background (Fig. 28); and many grasshoppers show similar changes. We shall consider such examples of morphological colour change in more detail when discussing polymorphism (p. 125). But it may well be that the neuroendocrine system is concerned in the production of these colour transformations. One well-documented example is the increased synthesis of ommochrome pigments which results in a brown patch which invariably appears overlying neurosecretory cells of Type B when these are taken from the suboesophageal ganglion of a black Phasmid and implanted into the green form.[158] And the brown coloration of *Pieris* pupae is brought about by a humoral factor released from the prothoracic ganglion and controlled by the brain through the oeso-phageal commissures.[148]

The epidermal cells of *Carausius* contain orange, red and yellow lipochromes, green biliverdin pigment, and varying amounts of the brown ommochrome pigments ommine and xanthommatine, which provide the various ' morphological ' colour types. And each form, except the green, has its own range of colours, brought about by the dispersion or clumping of the brown and orange pigments within the cells: a ' physiological ' colour change.[218] It has long been known that these reversible pigment movements depend on hormonal stimuli. Although the epidermal cells of *Carausius* are not innervated, if pieces of cuticle and epidermis are transplanted from one insect to another they change colour simultaneously with their new host.[218]

Carausius is normally dark at night and pale by day, and this change will continue for several weeks in complete darkness. This diurnal

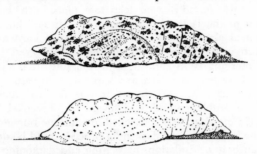

FIG. 28.   Effect of background on coloration of *Pieris* pupae. Above, darkly pigmented pupa formed on a black background; below, pure green pupa formed on a green or orange background.

rhythm is absent in insects kept in the dark from the time of hatching; it is induced by periodic illumination. Insects illuminated at night become pale; and a reversed rhythm[218] persisting in continuous darkness, can be induced by reversed illumination. Similar colour changes occur in Mantids.[218] The light stimulus is received solely through the eyes, for section of the eye stem eliminates the response;[218] but the details of the hormonal control of these colour changes are still not fully worked out. The neurosecretory cells located in the tritocerebrum of Phasmids, which are connected with the corpus cardiacum by the third corpus cardiacum nerve (Fig. 25), are probably involved, for an active substance can be extracted from this region of the brain which induces pigment movements.[158] Visible changes in the neurosecretory cells in the ventral nerve chain of Phasmids also take place in connection with the colour change—but no precise relation has been established.[158] A dual control from centres in the brain and in the corpus cardiacum seems also to regulate the expansion and contraction of the chromatophores on the air sacs of *Corethra*.[218]

Hardening and darkening, etc., of the cuticle

It has long seemed probable that the hardening and darkening of the cuticle which takes place simultaneously throughout the body shortly after moulting is under hormonal control. It was suggested that in *Rhodnius* the neurosecretory cells in the thoracic ganglia might be the source of the hormone in question;[213] but in unpublished experiments in which the abdomen was separated from the entire central nervous system, almost before the insect had escaped from the cast skin, by means of a transverse bar firmly pressed down just behind the metathorax, it was found that no delay took place in the hardening and pigmentation of the isolated abdomen. Clearly any hormonal factor responsible had already been released. Delayed hardening is not a normal feature in *Rhodnius*. C. B. Cottrell had observed that Lycaenid butterflies emerging from ants' nests in the soil often showed a prolonged delay between emergence and hardening; but since this was not convenient material for experiment it was decided that he should investigate the problem in Muscid flies. For Fraenkel in 1935 had shown that if *Calliphora* was compelled to continue burrowing through loose soil, the hardening and darkening process could be delayed for as long as seven hours. Inflation, hardening and darkening occurred as usual when the insects became free.

These chemical processes are evidently initiated by some nervous mechanism. Cottrell[41] found that the brain is concerned in the release to the blood of a peptide hormone which acts on the epidermal cells and leads to the normal darkening and hardening of the cuticle. It was shown by Fraenkel and Hsiao[56] that removal of the medial neurosecretory cells from the dorsum of the brain has the same effect as decapitation, suggesting that these are the source of the hormone; but they also noted that extracts from the compound ganglion of the thorax were even more active than extracts from the brain in inducing hardening in ligatured flies. The active substance begins to appear in the blood 2–3 minutes after emergence, reaches a maximum at 30–60 minutes and then falls and is quite inactive 10 hours later. These authors named the active substance ' bursicon ', a name that refers to the ' tanning ' of the cuticle after expansion. This name conveys the suggestion that the active substance tans the cuticle. That, of course, is not the case: the hormone is a messenger which stimulates the epidermis to carry out all the complex processes involved in hardening and darkening. In *Periplaneta*, ' bursicon ' occurs in most parts of the central nervous system, but it seems to be released only from the terminal abdominal ganglion. It appears in the blood soon after ecdysis and builds up to a maximum 1–2 hours later; it is no longer detectable at 8 hours. Section of the nerve cord stops release from the terminal ganglion; release can be induced by electrical stimulation.[133]

Another process which takes place at the time of moulting, and at eclosion from the egg, is the active absorption of fluid from the new tracheae and tracheoles, so that the system becomes filled with air. This likewise is a process for which the activity of the central nervous system is required: newly hatched mosquito larvae do not fill their tracheal system while lightly anaesthetized but do so soon after recovery. Here again a hormone is probably involved:[216] indeed, it might be ' bursicon ' itself.

Another change in the cuticle which appears to be under nervous control is the softening or ' plasticization ' which occurs in blood-sucking insects such as *Rhodnius* during the act of feeding and makes possible the rapid stretching of the cuticle by the large meal of blood, the result perhaps of a temporary increase in water content.[5] If a part of the nerve supply to the abdomen is cut, that region fails to expand and the insect after feeding has a lop-sided appearance. Examination of the abdominal nerves of *Rhodnius* with the electron microscope has revealed neurosecretory axons supplying the epidermis. This suggests that the epidermal cells of insects are under more localized endocrine control than had previously been supposed.[125]

Excretion and water balance

There has been some uncertainty about the nature of the stimulus to water secretion in insects. Injection of saline in large amounts into the body cavity of *Rhodnius* does not cause a rapid flow of urine although, after a meal of blood, there is copious excretion of water during the next three or four hours. From the observations of Maddrell[124] it appears that a circulating hormone is responsible; this becomes active within 2–3 minutes after feeding and accelerates secretion by the Malpighian tubules by more than a thousand times for a few hours. Isolated Malpighian tubules of *Rhodnius* immersed in haemolymph from an unfed *Rhodnius* show no signs of secretory activity; but haemolymph from a newly fed *Rhodnius* will induce a rapid flow of urine.

The factor responsible can be extracted from neurosecretory cells located in the posterior part of the fused ganglionic mass in the thorax, notably those in the metathorax. It can be extracted also from peripheral nerves, lying just behind the ganglionic mass, which contain axons from the neurosecretory cells. Certain of these axons send branches packed with neurosecretory granules to form swollen endings over the surface of the peripheral abdominal nerves shortly after they have left the ganglionic mass (Fig. 27). These are probably the sites of discharge of the diuretic hormone. The stimulus to secretion of the hormone comes via afferent nerves in the abdomen which respond to vertical stretching of the tergo-sternal muscles.[124] Of the various well-known neurohumours ' serotonin ' (5-hydroxytryptamine) is the only one which simulates the activity of the diuretic hormone in *Rhodnius*.[126] A similar hormone extractable chiefly from the medial neurosecretory cells of the brain has been found in the cotton stainer bug *Dysdercus*.

There is other evidence that neurosecretory cells may influence the water balance in the insect. In the larva of the aquatic beetle *Anisotarsus* extirpation of the dorsum of the brain and the corpus cardiacum upsets the water balance and water accumulates in the tissues; injection of saline extracts of brain and corpus cardiacum reduces this accumulation.[144] Likewise in *Schistocerca*,[76] *Locusta*[32] and *Gryllus*,[67] destruction of the neurosecretory cells of the pars intercerebralis of young adults causes distension of the abdomen from retention of water. Conversely the Malpighian tubules excrete more when the insect is injected with extracts from the pars intercerebralis of hydrated insects than from dehydrated insects, whereas rectal reabsorption is higher with extracts from dehydrated insects.[32] This latter regulatory process has been studied in *Periplaneta* which seems to have an ' antidiuretic hormone '; for saline extracts from brain and prothoracic ganglia of dehydrated *Periplaneta* will increase the reabsorption of water in the rectum. Moreover the isolated rectum of dehydrated animals absorbs water at a higher rate than that of hydrated animals.[206] A similar ' antidiuretic hormone ' can be extracted also from the perisympathetic organs of the abdomen in *Periplaneta*, Phasmids, etc.[9]

The four types of neurosecretory cells present in the ventral nerve chain of Phasmids show certain visible changes which seem to be connected with reproduction, changed water relations, changes in pigmentation, acceleration of the heart beat, etc., but their functions have not been worked out in detail.[159]

Neurohumours

The study of so-called ' neurohumours ' consists in the detection of pharmacologically active substances in extracts from the nervous system, or circulating in the haemolymph and presumably derived from the nervous system. Such substances commonly have a rather short-lived effect; they are eliminated rapidly in metabolism. They may be set free at a synapse within the central nervous system and serve to provide a chemical stimulus to the post-synaptic neurone; or they may be liberated at the neuromuscular junctions or other nerve endings, and serve to stimulate the muscle or gland cell concerned. These are the classic ' neurohumours '. But some of the materials to be discussed in this section are believed to be the products of neurosecretory cells and could therefore overlap with the hormones discussed in earlier sections of this chapter. There is no hard and fast line distinguishing ' neuro-humours ' from ' neurosecretions '.

The classic examples of neurohumours in vertebrates are adrenaline and nor-adrenaline which are set free at the endings of the sympathetic nervous system (and in larger amounts in the medulla of the suprarenal gland) and acetylcholine which is liberated at the endings of the parasympathetic nervous system (such as the vagus nerve). In both systems the active substance exerts its effect not only locally upon the muscle supplied or the nerve axon contacted, but is set free into the circulating blood: the blood flowing from a heart slowed by action of the vagus nerve will slow the beating of another heart to which it is applied. Acetylcholine is highly important also as the chemical trans-mitter between one nerve and the next in the synapses of the central nervous system and as the chemical transmitter at the end-plates or neuromuscular junctions in voluntary muscles.

Catecholamines

Insects contain relatively large amounts of catecholamines, notably 3-hydroxytyramine (dopamine),[201] while nor-adrenaline and adrenaline occur in smaller quantities.[201] Adrenaline, nor-adrenaline and dopa-mine produce neurological effects in the cockroach: increased activity in the central nervous system, and facilitation followed by blocking of the giant fibre synapses on the tract from the cercal nerve.[81] A charac-teristic yellow-green fluorescence can be utilized to demonstrate them histologically. In the brain of *Periplaneta* there are distinct regions where the nerve cells show this property; and this has led to the sug-gestion that some catecholamine acts as a synaptic transmitter. No evidence could be obtained by this means of the presence of cate-cholamines in the corpus cardiacum.[57] There is evidence also that an adrenaline-like transmitter operates peripherally in the luminous organ of the fire-fly *Photuris*. Thus reserpine (a drug that is known to drain vertebrate nerve endings of adrenergic transmitters) abolishes the luminescence produced by nerve stimulation.[29] Certainly adrenaline will produce pharmacological effects on the heart beat, gut movements, etc., in insects.

Acetylcholine

Acetylcholine is present in large amounts in the central nervous system of insects;[218] the nerve cord of *Periplaneta* contains some fifteen times the average concentration in the mammalian central nervous system,[218] and cholinesterase, the specific enzyme which hydrolyses and inactivates acetylcholine, exists in very high concentration in the nervous tissues of cockroach, honey-bee and other insects.[218] The enzyme is concentrated in the neuropile, particularly where synaptic sites are most numerous; it is also present in the glial membranes at the periphery of the ganglion of *Periplaneta* and even in the fibrous sheath, the neural lamella, so that the central nervous system is well shielded from the entry of acetylcholine from outside.[202] Every effort seems to be made by the organism to localize it and eliminate it at the synaptic site. In *Oncopeltus, Musca*, etc., choline acetylase (concerned in the synthesis of acetylcholine) first becomes detectable in the embryo at the time of appearance of the neuroblasts.[39, 201] Injection of esserine (a specific inhibitor of acetylcholinesterase) into the fire-fly *Photuris* will induce rapid irregular flashing, an action that takes place in the ganglia which supply the luminous organ.[29]

There is thus strong circumstantial evidence that acetylcholine serves as a transmitter substance in the neuropile.[218] It is sometimes claimed that the presynaptic vesicles, which closely resemble the electronlucent type of neurosecretory granules (p. 89), are the carriers from which acetylcholine is liberated. Acetylcholine at a very low concentration will cause an increased rate of discharge in isolated cardiac ganglion cells in *Periplaneta*. This action does not seem to be at a synapse, as is believed to occur in the central nervous system, but is perhaps an action on the ganglion cells themselves.[131] Acetylcholine is certainly not the chemical transmitter at the neuromuscular junctions, in striking contrast to the muscles of vertebrates; cholinesterase is absent at the axon terminations in insect muscle.[218]

Other neurohumours

5-Hydroxytryptamine (serotonin) is readily extractable from the central nervous system of insects and from the corpus cardiacum.[39, 201] It appears to come from certain of the neurosecretory cells.[80] It has been described as having an excitatory effect on the gut and on the heart in *Periplaneta*,[40] as having either a stimulating or an inhibiting action in neuromuscular transmission[201] and as raising or lowering the threshold of activation in the central nervous system of Lepidoptera in which it enhances night flight.[80] In *Calliphora* at very low doses it will increase the rate of fluid secretion by isolated salivary glands.[7] We have already seen that 5-hydroxytryptamine will simulate the action of the diuretic hormone in *Rhodnius* (p. 100).[126]

γ-Amino butyric acid (GABA) which may function as a chemical transmitter during inhibition in vertebrates, has been shown to cause inhibition of impulse conduction in the caterpillar of *Dendrolimus*.[201] γ-Amino butyric acid is produced in metabolism by decarboxylation of glutamic acid, the decarboxylase in question being plentiful in insect nervous tissue.[201]   It appears likely that the parent amino acid is itself important as a transmitter substance, notably at the neuromuscular junctions.   Thus L-glutamate applied iontophoretically to muscle fibres of *Schistocerca* and *Tenebrio* induces depolarization at the synaptic sites; and the L-glutamate that is liberated from muscle increases during stimulation of the motor nerve supply.[204]

In addition to the foregoing, there are other pharmacologically active substances to be found in the tissues of insects.   A material differing from all the above-named substances is extractable from insect nerves and will induce slow contractions in the mid-gut muscles.[20] Some six different substances can be extracted from the corpus cardiacum-allatum complex of *Periplaneta* which accelerate the beating of the isolated heart; and some further substances which slow the heart beat have been extracted from the central nervous system.[161]   How far such materials are of physiological significance and how far they are products of tissue breakdown is uncertain.

We have already mentioned the material contained in the opaque accessory gland secretion in male insects which acts in stimulating the nerves that control peristalsis in the female reproductive tract.[43]   This seems to be a pharmacologically active amine which also stimulates the heart beat and resembles the active material liberated apparently from the pericardial cells of *Periplaneta* under the action of a neurosecretory peptide from the corpus cardiacum.[43, 97]

6: Metabolic Hormones and Feed-back Effects

All hormones exert their effects by influencing metabolism. The regulation of moulting and of diapause, the control of metamorphosis and reproduction, the influence of neurosecretory products on diurnal rhythms, colour change, cuticle hardening, or excretion and water balance, which have been the subject of earlier chapters, can all be viewed in that way; as can also many of the short-term effects of neurohumours.

The action of many hormones results in the unmasking of some latent component in the gene system which evokes the most profound changes in metabolism. We have considered this effect in relation to metamorphosis and we shall return to it in discussing differentiation and polymorphism (Chapter 7). In this chapter we shall be concerned mainly with trying to distinguish between metabolic effects and feed-back effects, or ' homeostasis '.

It is by no means easy to distinguish between direct effects of hormones on some specific process, and a change in metabolism which is a consequence of demands elsewhere in the body; demands which are themselves the result of hormonal action. That is what is meant by feed-back effects. It is certain that both these processes occur; but it is perhaps a good rule to assume that observed changes in metabolism are feed-back effects unless that can be excluded by experiment.

We considered this problem in relation to the fall in oxygen consumption in diapause and in the young pupa, and concluded that this fall is generally due to the cessation of the endothermic syntheses that are necessary for growth and reproduction; and that it is the processes of growth which are set going by the hormones (p. 18). But this may not be the whole story: extirpation of the corpora allata in the adult female *Leptinotarsa* induces the diapause state, with a reduced level of oxygen consumption. To a great extent that fall must result from the arrest of egg production. But the oxygen consumption of tissue homogenates from diapausing *Leptinotarsa* is increased by the addition of active corpora allata or juvenile hormone, which may thus have an additional direct action at the subcellular level in stimulating the activity of respiratory enzymes.[190]

Corpus allatum: metabolic action and feed-back in maturation of the ovaries

In *Rhodnius* females deprived of their corpus allatum by decapitation the oöcytes fail to develop yolk; and the large meal of blood in the stomach is digested very slowly. (The fat body in *Rhodnius* is relatively inconspicuous; it is important in intermediary metabolism, but the main reserve for growth is the undigested blood in the stomach.) If the fore part of the head, containing the brain and neurosecretory cells, is cut away, leaving the corpus allatum (and corpus cardiacum) intact, the blood meal is rapidly digested and the full complement of eggs is developed. The accelerated digestion in the presence of the corpus allatum is perhaps a feed-back effect from the developing ovary.[216] This conclusion is supported by the fact that if the ovaries are removed from female larvae in the fourth stage, so that castrated adult females are obtained which have completely recovered from the injury of the operation, there is the same delay in the digestion of blood; and this happens whether the corpus allatum is retained or not.[216]

Similar effects can be seen in *Drosophila*. In the mutant called ' female sterile adipose ' (*adp/adp*) the sterility is due to an autonomous defect in the ovaries: if wild-type ovaries are implanted into this mutant they develop normally. But whereas the fat body of *adp/adp* females ordinarily stores excessive quantities of fat, it does not do so if wild-type ovaries are implanted. The accumulation of fat is clearly a homeostatic response.[46]

In the adult female of *Pyrrhocoris apterus* the protein content of the haemolymph increases during the reproductive cycle (protein concentration in the male is constantly low). But in both castrated and allatectomized females the protein concentration is nearly four times higher than in the normal female. If corpora allata are implanted in the allatectomized female ovarian development is restored and haemolymph protein falls to normal levels. The effect of the corpus allatum is therefore indirect—but digestion in the gut and protein synthesis in the fat body are probably a direct metabolic effect of the neurosecretory hormone from the brain (p. 106).[181]

Allatectomy in *Periplaneta* slows down the turnover of both triglyceride and phospholipid. But here again the corpus allatum appears to regulate metabolism of these materials by controlling their utilization. The greater accumulation of fats in the fat body after removal of the corpus allatum is due to the failure of ovarian development, since ovarian lipid is 70 per cent triglyceride.[205] Likewise in *Leucophaea*, the ovary can undoubtedly synthesize triglycerides; and this process is enhanced under the action of the juvenile hormone, whereas lipid synthesis in the fat body is depressed and thus more substance is made available for the developing oöcytes.[63] These results could again be described as due to feed-back effects.

Some of the clearest evidence of a direct effect of the corpus allatum on metabolic processes outside the ovary was given by Pfeiffer.[216] In each maturation period in the female grasshopper *Melanoplus* there is a great increase in the size of the fat body. At the close of this period yolk production begins and hypertrophy of the fat body ceases. This change in fat body metabolism, however, is not caused by the ovaries, for it occurs also in castrated females which retain their corpora allata. But if the corpora allata are removed the fat body continues to enlarge and becomes enormously hypertrophied. The corpora allata seem to affect fat metabolism in this insect in two ways (*a*) by inducing utilization of stored fat in a process independent of the ovaries; (*b*) by facilitating production of yolk, which in turn affects fat metabolism.

A different type of feed-back effect was suggested by Odhiambo[146] to explain the failure of ovariectomy to lead to the accumulation of reserves in the fat body of locusts and grasshoppers. He noted that in *Schistocerca* males removal of the corpora allata leads to persistent inactivity and suggested that it is this sedentary behaviour, due presumably to an action of the juvenile hormone on the central nervous system, which leads to the build-up of glycogen and fat. In studying *Locusta*, Strong[196] came to a somewhat different conclusion, namely that, as suggested by Pfeiffer in *Melanoplus*, the corpus allatum causes mobilization of fat-body reserves and this mobilization makes possible the continuous activity of these locusts which ensures successful mating.

Neurosecretory cells and protein synthesis

The neurosecretory cells in the *pars intercerebralis* play an important role in the maturation of the ovaries in many insects (p. 77). In general terms it would seem that whereas the action of the corpus allatum secretion is exerted mainly upon the ovaries themselves, the product of the neurosecretory cells has largely a metabolic effect. It was observed by E. Thomsen[212] that in the adult female of *Calliphora*, if the neurosecretory cells are excised from the dorsum of the brain, the fat body becomes stuffed with glycogen, and yolk fails to develop in the oöcytes. The eggs reach the same stage of development as in flies fed on sugar alone. It would appear that the basic deficiency is in the production of the protein needed for yolk formation. An important effect of the hormone is to stimulate the production of proteolytic enzymes in the mid-gut of the fly that has been fed on meat and thus to furnish the amino acids needed for protein synthesis in the fat body. Protease activity in females deprived of the medial neurosecretory cells is only about one-quarter to one-third of that in normal females on the same diet.[200] Since proteases are themselves proteins, Thomsen and Møller suggest that the hormone of the neurosecretory cells may be involved in protein synthesis in general. These cells are also responsible for the development of esterase in the cells of the mid-gut and for this development a protein-rich diet is again necessary.[200]

The importance of the hormone from the medial neurosecretory cells in the protein metabolism of locusts has been made very clear by by the work of Hill[78] and Highnam et al.[216] Whereas in Schistocerca the corpus allatum hormone appears to act directly upon the growing oöcytes and the follicular cells of the ovary, the neurosecretory system acts upon the protein metabolism of the body. When the neurosecretory system releases its secretion into the haemolymph, the blood protein rises; when the secretory material is accumulating in the system, the blood protein falls. The neurosecretory product induces the fat body to synthesize blood protein from the circulating amino acids; if isotopically labelled glycine is injected it appears in the form of protein, first of all in the fat body cells, then in the haemolymph, and finally it begins to appear in the oöcytes. It is the formed proteins of the haemolymph which are transferred by the follicular cells into the developing oöcyte (p. 81). When progressively larger amounts of the ovary were removed from a series of Schistocerca females the concentration of protein in the haemolymph rose from around 3 per cent in normal females, to 7·3 per cent when two-thirds of the ovaries had been removed, and to 13·4 per cent when ovariectomy was nearly complete. Conversely, cauterization of the neurosecretory cells results in a low level of haemolymph protein (1–1·5 per cent), whereas all those factors which induce liberation of neurosecretory material and accelerate maturation of the ovaries (hyperactivity, rearing alongside mature males, copulation, etc. (p. 82)) also increase the level of haemolymph protein.

In Schistocerca, as in many other insects (p. 72) the removal of the corpus allatum leads to the resorption of oöcytes. This is the result of (i) competition between oöcytes for juvenile hormone and (ii) competition for available protein, the product of neurosecretory activity. The corpora allata appear to exert an activating effect upon the neurosecretory cells: if they are removed in Schistocerca the neurosecretory system becomes inactive. It may be that the observed decrease in oxygen consumption which follows removal of the corpus allatum may be due to the resulting cessation of protein synthesis and not to any direct effect of the corpus allatum hormone.[216] These conclusions have in general been supported by the results of Minks[133a] on *Locusta*, but he suggests that whereas the neurosecretory cells activate protein synthesis in the fat body, the juvenile hormone exerts its effect in the presence of this active protein synthesis and induces the formation of specific vitellogenic proteins. Coles[38] likewise formed the opinion that the corpus allatum secretion in *Rhodnius* has a specific action on the fat body, inducing the formation of specific ' yolk proteins ' ready for transfer to the oöcytes: in the castrated female ' yolk proteins ' accumulate in the haemolymph, but only if the corpus allatum is retained. Similar conclusions have been reached in *Locusta*[194] and in *Leucophaea*.[50] But in these insects and in *Schistocerca*[78] the interrelations between the corpora allata and the cerebral neurosecretory cells are not yet clear.[44]

Homeostasis and humoral integration

The foregoing evidence strongly supports the belief that insect hormones can act directly upon such tissues as the fat body or the gut wall, which are concerned in intermediary metabolism, and thus produce the materials that are needed for the growth processes that the hormones have set going. But, as always, it is necessary to exclude the possibility that these metabolic effects are homeostatic responses. When the neurosecretory material in *Calliphora* females causes increased protease secretion in the mid-gut, is it certain that this is not a feed-back response, via the haemolymph, from the fat body which is demanding amino acids for protein synthesis? And is it certain that the synthetic activity of the fat body is not a feed-back response from the developing ovaries which are demanding protein? It is probable that both processes are operating at the same time and that the relative importance of the two processes (direct action and feed-back effect) differ in different insects. In *Schistocerca*, as reviewed above, the experiments make it clear that the neurosecretory cells are stimulating protein synthesis whether the ovaries are present or not.

Another example of a metabolic effect in which feed-back responses seem to be excluded is the factor extracted by Steele[188] from the corpus cardiacum of *Periplaneta*. This extract contains a powerful ' hyperglycaemic factor ' which leads to an increased level of blood trehalose with a simultaneous decline of fat-body glycogen. Its effect appears to depend on its ability to increase the activity of the enzyme phosphorylase. In contrast with this the diapause hormone of the silkworm pupa

(p. 40) which has been available as an active extract, induces a fall in fat-body glycogen and in the trehalose of the circulating haemolymph; it appears to be a ' hypoglycaemic factor '. Actually it exerts its effect directly on the pupal ovaries where it stimulates the enzyme system responsible for glycogen synthesis; the action on the fat body and the blood of the pupa is a feed-back effect. Perhaps the primary site of action is the gene system in the ovarian follicle cells.[73]

The existence of feed-back responses in the physiological working of the body is accepted as a fact of life and in the insect at least such responses have not been subjected to much detailed analysis. Such interchange and mutual communication within the body is of prime importance at all stages of growth; we shall be concerned with these relations in the next chapter. But such exchanges are of equal importance in the day to day control of the chemistry of the body. This chemistry conforms to a standard for each part of the body which, given appropriate nutrition, is kept within fairly well-defined limits. The circulating haemolymph is the common pool from which all the tissues can help themselves. Any deficiency or excess within this pool is corrected in the short term by the release of material from storage, or by synthesis, on the one hand, and by excretion, breakdown or consignment to reserve, on the other. Some defects are recognized by the nervous system and corrected by nervous means—as deficiency of oxygen or excess of carbon dioxide, or acid metabolites, are countered by increased respiration. Others may be corrected by hormones which control specific metabolic changes. Many departures from the normal must presumably be recognized and corrected by the peripheral cells and tissues themselves.

These are the processes of homeostasis which are central to all physiology but about which we are singularly ill informed.

7: Hormones in Differentiation, Regeneration and Polymorphism

In considering the nature of metamorphosis (Chapter 3) we reached the conclusion that it represents the delayed manifestation of characters that remain latent, or ' coded ', within the gene system during the growing stages. The present chapter deals with the very similar problems that arise in differentiation, regeneration and polymorphism; but we shall be concerned with these topics only in so far as there is evidence of chemical or hormonal factors being involved.

The cellular differences that are acquired during embryonic development, during regeneration of an organ, or when nuclei are exposed to different cytoplasmic environments, are usually regarded today as resulting from alterations of gene activity in the nuclei. The genes are not considered to be changed, but merely to be activated, or derepressed. It is possible to picture this as a chain reaction: an initial stimulus may cause changes which prepare the system to respond to consequent stimuli by further changes, and so on. By methods to be discussed these changes are distributed in a pattern which is the basis of the body form.

Embryonic development

At the time when embryonic development begins the oöplasm has already become organized into a pattern of regionally distinct properties. This pattern is for the most part invisible, although in some insects, such as *Tenebrio*, the region of the cortical plasma which will be the site of the future germ band, stains more deeply. The existence of such a pattern is proved by the fate of the nuclei which enter its different parts. According to the principle discovered by Boveri in the later years of the last century it is the cytoplasm which decides the fate of these nuclei. Clear evidence of this was provided by Hegner in 1908, who showed that those cleavage nuclei which enter the cortical plasma of the posterior pole of the egg in *Leptinotarsa* form recognizably distinct ' pole cells ' which ultimately become the germ cells of the gonads.[218] In *Drosophila* a cleavage nucleus may divide so that one daughter nucleus lies in the polar plasma, the other outside it; then the former will become a germ cell, the latter an ordinary blastoderm cell.[218] The cleavage nuclei are clearly subject to a ' chemo-differentiation ' which exists in the cortical plasma.

The entire cortical cytoplasm of the egg is rich in nucleoprotein, but the relative importance of protein and nucleic acid in the early stages of determination is uncertain.[218] Ribonucleic acid is particularly evident in the cytoplasm of the posterior pole, the so-called ' oösome ', which, in the unfertilized egg of *Apanteles* is a prominent site of alkaline phosphatase activity.[218] In *Drosophila* the polar granules are made up of dense masses of ribosomes[218] which are later included within the pole cells.[218] In Cecidomyid midges the germ-line nuclei retain their full complement of chromosomes, whereas most of the chromosomes are lost from the future somatic nuclei. But if the somatic nuclei are exposed to the nucleoprotein of the pole plasma, the elimination of chromosomes is arrested.[218] It may be that the characteristic effects, which the different regions of the cortical plasma exert upon the nuclei that enter them, may depend upon differences in the nature of the RNA content of each zone. But it is when the nuclei establish themselves in the peripheral cytoplasm that RNA synthesis becomes conspicuous.[115]

The foregoing account relates to ' mosaic ' eggs of Diptera, etc., in which the general pattern of the cortical plasma is determined before the cleavage nuclei arrive. In many developing eggs, for example in the dragonfly *Platycnemis*, as studied by Seidel, determination of the main regions of the embryo does not take place until some time after the cleavage nuclei have reached the cortical plasma.[218] Such eggs are capable of extensive ' regulation ' after they have been laid. If a substantial part of the anterior region of the egg is removed by ligature, a well-proportioned but reduced embryo is differentiated in the remaining compartment of the egg. If the germ band is divided longitudinally, two well-proportioned embryos are formed.[102]

The arrival of the cleavage nuclei at the posterior pole of the egg of *Platycnemis* sets in train the whole process of embryo development. If this quite restricted region is eliminated at a very early stage by excision or by exposure to ultra-violet light, cleavage and migration of nuclei occur as usual, but the resulting blastoderm is solely of the extra-embryonic type; no germ band is formed.[218] Elimination of other parts of the egg do not have this effect. A centre of this type has been found in *Platycnemis*, the ant *Camponotus*, the beetles *Sitona*, *Bruchus* and

Tenebrio, and in the mosquito *Culex*. There are other insects in which
it has not been found: *Leptinotarsa, Melasoma* (Col.), *Calliphora*. In the
leafhopper *Euscelis* the centre in question is closely associated with the

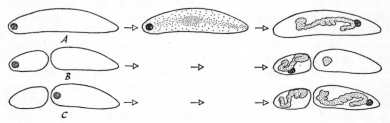

FIG. 29.   The effect of posterior pole material on the induction of
embryonic development in the egg of *Euscelis* (after Sander).   A,
normal development; the dark sphere represents the ball of symbionts.
B, effect of ligature at an early stage; an embryo is developed only in the
posterior fragment.   C, the ball of symbionts with associated cytoplasm, is displaced forewards before ligature; an embryo is developed
in both fragments.

mass of symbionts transmitted through the egg.   If this mass is displaced experimentally into the anterior half of the egg a complete
embryo is induced in this region[102] (Fig. 29).   If the eggs of *Chironomus*
are centrifuged strange abnormalities may arise, such as larvae with a
head at each end and no tail; or larvae with a tail at each end and no

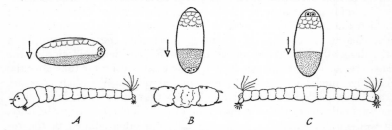

FIG. 30.   The effect of centrifugation on development of the embryo in
the egg of *Chironomus* (after Yajima)   A, normal development; B, an
embryo with two heads; C, an embryo with two abdomens.

head[153] (Fig. 30).   Presumably the activation centre is involved in this
response and can be shifted by centrifugation.

This centre at the posterior pole is named the ' activation centre '.
It is believed to exert its effect by giving off a material substance which
permeates the egg from behind forward.[102]   This is the nearest thing to
an endocrine centre that has been found in the early embryo.   A second
centre has been described in the eggs of most insects, localized usually

towards the middle of the presumptive germ band, in a position corresponding with the future thorax of the embryo. This is called the ' differentiation centre '. All the successive changes in early embryonic development seem to take place first in this region and to spread forward and backward. There is no evidence that the differentiation centre is giving off a diffusible substance. It seems to be the central point of a developmental gradient the nature of which is not known.

During the later stages of differentiation, when the mesoderm and endoderm are being formed, it appears that, contrary to the situation in vertebrates, the ectoderm is the self-differentiating system, independent of the mesoderm. Indeed the differentiation of mesodermal structures in *Chrysopa* and in *Leptinotarsa* depends upon the overlying ectoderm, which seems to induce the corresponding mesodermal organs in much the same way as the mesoderm in Amphibia induces differentiation in the ectoderm. Whether these changes are brought about by chemical (hormonal) communication in some form is unknown.[102, 218]

In the final stages of embryonic development many insects undergo an embryonic moult; sometimes two embryonic moults; the final cuticle being shed during the act of eclosion from the egg. By that time all the organs and tissues are fully differentiated and it might be expected that this moulting process would be controlled in the usual way by neurosecretory cells and prothoracic glands; but in *Melanoplus* both embryonic moults seem to be entirely independent of nervous and endocrine centres in the head, thorax, or abdomen.[136] The same applies to the first embryonic moult in Gryllid eggs.[69]

Differentiation in the post-embryonic epidermis

The influence of the cell plasma on the fate of nuclei is well seen in the neuroblasts of the grasshopper embryo. By repeated division these cells give rise to a succession of ganglion cells. It was shown by Carlson[30] that when these cells are already in metaphase it is possible, with a microdissection needle, to rotate the mitotic spindle through 180° so that the chromosomes which would have remained in the neuroblast how go to the ganglion cell. But this makes no difference to the result of the division: it is the cytoplasm and not the chromosomes that determine which cell shall be the ganglion cell and which the continuing neuroblast.

A similar ' differentiative division ' of a cell is to be seen in the formation of the sense organs and other ' organules ' in the insect integument (Fig. 31). A tactile hair is derived from a single cell which divides into four cells that become respectively the hair-forming, or trichogen cell, the socket-forming, or ' tormogen cell ', the sense cell

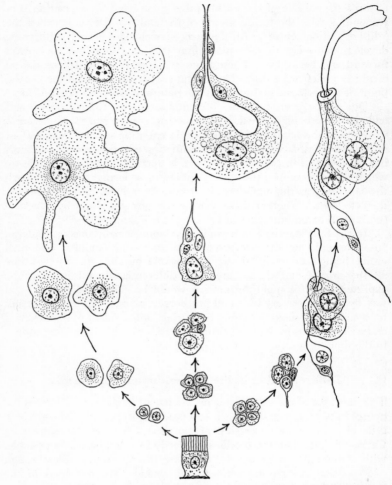

FIG. 31. Three types of differentiation available to the epidermal cell in *Rhodnius*. To the left, formation of oenocytes; in the centre, formation of dermal gland; to the right formation of tactile sense organ.

which sends a distal process to the base of the hair and a proximal process, or axon, which grows inwards to reach ultimately the central nervous system, and a neurilemma cell, or Schwann cell, which provides a sheath for the sense cell and axon.[218] The determination of the future fate of each of these cells is presumably brought about by chemical differences in those regions of the cytoplasm in which their nuclei come to lie. Once more the gene system in these nuclei is presumably trans-determined, or ' programmed ', by chemical factors in the cytoplasm. The distribution of these chemical factors may in turn be controlled by the immediate environment of the parent cell. And one important element in that environment will be the position it occupies in the various gradients that are operating in the integument.[105]

There is evidence for the existence of gradients of some kind in the abdominal segments of insects and in the segments of the legs.[114] In the milkweed bug *Oncopeltus* the adult cuticle of the abdomen is characterized by a great number of small non-innervated hairs which make their appearance at metamorphosis. It was shown by Lawrence[105] that during their formation these hairs orient themselves to some unseen influence in the epidermis. The observed results can be described by the hypothesis that the epidermal cells create a chemical gradient by ' pumping ' in an antero-posterior direction and that it is this actively maintained gradient which informs the hair-forming cells and thus compels them to conform to the polarity of the abdomen as a whole.

An elaboration of this interpretation suggests that a particular structure, that is, the type of cuticle laid down, may be controlled by the absolute ' level ' in the gradient at which the cells find themselves.[105, 197] After wounding, the cells may temporarily lose their polarity (just as they lose the capacity to form cuticle of a particular kind) but later, under the influence of the neighbouring cells, the capacities are recovered.[105, 113]

There is a general belief that 'patterns' of structure, pigmentation, etc., are controlled by more or less diffusible substances, which regulate gene activity; but it has proved very difficult to obtain good experimental evidence for this idea. A suggestive example is seen in the external male genitalia of *Drosophila*. The anal plate carries a region with ' teeth ' and a region with ' bristles '. During normal development, and also after the transplantation of primordia into another insect, it is often possible to see structures of an intermediate or composite type in the zone between the two regions, suggesting the existence of a diffusible substance whose concentration decreases in a gradient between the two zones.[119]

A somewhat different gradient hypothesis has been put forward to account for the distribution of tactile hairs and dermal glands in the integument. Over the abdomen in *Rhodnius* larvae the tactile hairs arise at the centre of plaques which are fairly evenly distributed over the surface (Fig. 32A). The dermal glands, which likewise arise by the division of one cell into four daughter cells (Fig. 31), are more numerous than the secretory hairs and are therefore less widely spaced. At each moult in the young stages of *Rhodnius* new sensory hairs and new dermal glands arise from the ordinary epidermal cells by differentiative division as already described. The sensory hairs appear at those points where the existing hairs are most widely separated (Fig. 32B). They never appear in close contact with an existing hair. There seems little doubt that the existing hairs inhibit the appearance of new hairs in their immediate vicinity. If we really understood the nature of this inhibition

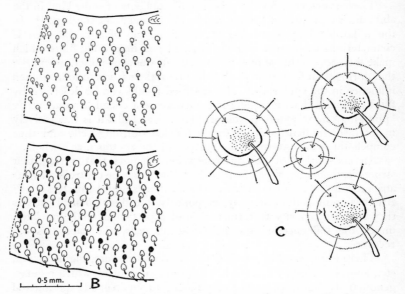

FIG. 32. A, part of the third abdominal tergite in normal 4th-stage larva of *Rhodnius*. B, the same area after moulting to the 5th stage, with newly differentiated plaques black. C, diagram to illustrate hypothesis of determination of new sensory hairs. Explanation in text.

we should have come a long way towards understanding the process of differentiation.

According to the interpretation outlined in this chapter, the transforming factor, which will evoke in the ordinary epidermal cell those components of the gene system that lead to the development of a sensory hair, is already present in the plasma of the epidermal cells. A developing 'sensillum' is pictured as taking up this hypothetical inductor, or transforming factor, and so draining it away from the surrounding cells (Fig. 32C). These cells are thereby deprived of the transforming factor and thus the formation of other sensory hairs in the immediate neighbourhood is inhibited.

If an area of the integument is killed by burning, the epidermis is restored by the multiplication and ingrowth of epidermal cells around the margin of the burn.   If such a burn is inflicted on a *Rhodnius* larva shortly after feeding, that is, just when growth and moulting are beginning, the new integument that is formed at the next moult carries no sensory hairs (Fig. 33B): these do not appear until the following moult.   But dermal glands have been differentiated in the usual

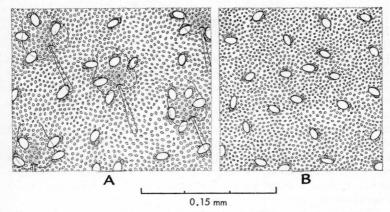

0.15 mm

FIG. 33.   A, epidermis of 4th-stage larva of *Rhodnius* mounted shortly before moulting to the 5th stage, showing the distribution of the plaques and sensory hairs and the distended dermal glands.   B, the epidermis of the same insect over an area regenerated after a burn.   Dermal glands have been differentiated, but no sensory hairs.

numbers and at the normal separation.   This result suggests that perhaps the dermal glands are induced by the same transforming agent as that needed for the determination of sensory hairs, but at a lower concentration.[217]   This conception, that quantitative differences in the amount or concentration of a transforming agent may induce qualitative differences in the transformation that is brought about (such as the juvenile hormone does (p. 63)) was an essential part of the gradient theory of differentiation developed many years ago by E. C. Child. Unfortunately there is no direct evidence for the existence of these transforming agents in *Rhodnius*; they have merely been postulated to account for the experimental results.   But on the basis of this interpretation it is possible to build up a conception of differentiation applicable to the body as a whole.[217]

It has often been suggested that the periodic distribution of structures or pigment patterns in the integument of insects may have a basis in the Liesegang phenomenon of periodic precipitation.[217]   Sometimes

hairs and scales are arranged in linear fashion, as are scales on the wings
of Lepidoptera or bristles on the legs of *Drosophila*, and Hollingsworth[82]
has formulated a model to describe how this could be accounted for
along these lines.

Differentiation and cell death

Cell death is a universal element in all growth and development (p. 66).
During postembryonic development in insects the evidence of cell death
is commonly seen in the epidermis in the form of ' chromatic droplets '.
These droplets which contain RNA and DNA are derived from the
nuclei and cytoplasm of autolysed cells.   Some of the cells that die in
this way appear to be cells that are surplus to requirement: mitosis seems
to be so vigorous during moulting that more cells are produced than are
needed for the new integument.   But this process is most conspicuous
at metamorphosis when hairs, dermal glands and other structures are
not maintained in the adult stage and their formative cells die and
disintegrate.[215] Scales in Lepidoptera are homologous with tactile hairs;
but they are not innervated and the cell that would provide the sense
cell and neurilemma cell undergoes autolysis.   The important example
of the prothoracic gland and the hormonal mechanism which leads to its
breakdown at metamorphosis has already been described (p. 66).

The insect never allows an organ that is no longer physiologically
active to hold on to reserves of protein which could be more usefully
employed elsewhere.   In *Rhodnius* the intersegmental muscles of the
abdominal sternites are functional only at the time of ecdysis when
contraction is needed to provide the hydrostatic pressure in the haemolymph which will help to rupture the old skin and drive the blood into
the wings and legs to bring about their expansion.   Since *Rhodnius*
ingests an enormous quantity of blood at its one meal in each stadium
it will be a positive advantage not to have the muscles at that time.
Indeed, as soon as the moult is over these muscles undergo a rapid
autolysis, the products being used, apparently, to build up the inner
layers of the new cuticle.   In this case the nuclei and muscle sheath
survive; the contractile fibres and the mitochondria are almost completely broken down and translocated (Fig. 34).   The humoral control
of this process is not known; it is independent of the nerve supply.[218]
Likewise in the *Calliphora* adult, within a few days after emergence
many of the muscles concerned in the production of digging movements
and the expansion of the cuticle degenerate.[41]

In other insects the massive flight muscles are an important source
of protein for the production of eggs.   In the queen ant they are used
for the mating flight; but then the female sheds her wings and the

muscle protein is transferred to the ovaries.   The same change is seen in many Aphids after their sole migratory flight[218] and even in the mosquitoes of northern latitudes where a blood supply is hard to come by.[218]   The nature of the hormonal control of these muscle changes is almost unknown.   We saw that the reversible mobilization of the flight muscle protein of *Leptinotarsa* that occurs in adult diapause (p. 85) is regulated by juvenile hormone secretion.   The breakdown of muscle in

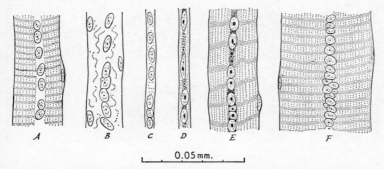

0.05 mm.

FIG. 34.   Cycle of involution and new formation of the ventral abdominal muscles in the 1st-stage larva of *Rhodnius*.   A, immediately after hatching. B, 4 days after hatching.   C, immediately after feeding (14 days after hatching).   D, 4 days after feeding, new muscle fibrils beginning to appear.   E, 7 days after feeding.   F, 11 days after feeding (newly moulted 2nd-stage larva).

Saturniid moths is naturally prevented by juvenile hormone injection because this results in the retention of many of the pupal characters.[116]

Wound healing

Insects in all stages can repair injuries to their integument.   The epidermis plays the chief part in this response (p. 56).   As studied in *Rhodnius* it appears that the dead or injured cells quickly release substances, products perhaps of the hydrolysis of proteins (since proteoses and peptones are active in this respect) which exert an attraction upon the surrounding cells so that these migrate to the wound to congregate thickly around its margin, leaving a peripheral zone where the epidermal cells are very sparse (Fig. 35).   If a piece of the integument has been removed the aggregated cells spread across the wound, make good the defect, and lay down a new cuticle of normal character.   At the margins of the wound the new cuticle extends for some distance under the old. Meanwhile cell divisions take place in the sparse peripheral zone and continue until the normal density of the cells has been restored.[217]   The process of healing follows essentially the same course in larvae of *Ephestia*[218] and *Locusta*.[165]

Wound healing takes place similarly in the decapitated *Rhodnius* larva and in the adult. It is clearly not dependent on the central endocrine system. But the cytological changes, the activation of the nuclei with enlargement of the nucleolus, the rapid accumulation of

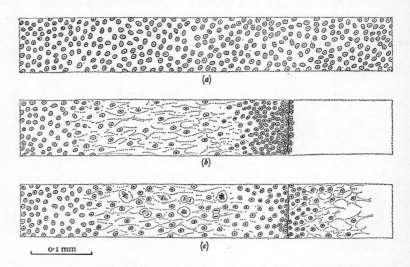

FIG. 35. (a) Surface view of epidermis in normal adult *Rhodnius*. (b) The same 24 hours after an excision about 1 mm square. The cells are crowded along the cut margin. (c) The same 4 days after the excision. Cells are spreading over the excised area.

ribonucleic acid and new protein in the cytoplasm, with a great increase in the number of mitochondria, are identical with those seen in the epidermal cells when they are induced to grow under the action of ecdysone.[215] During this new cuticle formation there is no local metamorphosis in the epidermal cells even in the absence of juvenile hormone: a wound in the diapausing pupa of *Hyalophora* is repaired with pupal and not with adult cuticle.[8]

The chemical stimuli from injured cells are sometimes referred to as 'wound hormones' but their nature has not been clearly established. In diapausing larvae of *Lucilia* the arrest of growth may be brought to an end by pricking or singeing. That does not happen in decapitated *Rhodnius*, nor in the diapausing pupa of *Hyalophora*. In *Hyalophora* the products of injury have a fairly widespread effect in increasing the number of circulating haemocytes and raising the oxygen consumption, but there is no general renewal of growth.[227]

So far as the local response is concerned the initial aggregation of the surrounding cells seems to depend on ' attraction ' by the injured cells. But the spreading over the wound seems to result from lack of contact with the missing cells.   As the cells migrate they always maintain contact with their surviving neighbours, even though they may do so

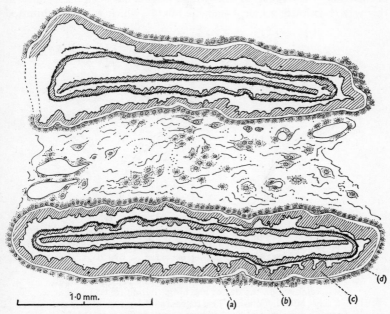

1·0 mm.

(a)   (b)   (c)   (d)

FIG. 36.   Longitudinal section of fragment of tibia of adult *Rhodnius* implanted in 4th-stage larva and sectioned when this host became adult. (a) Original cuticle; (b) first new cuticle forming continuous capsule; (c) second new cuticle forming continuous capsule; (d) epidermis.

only by tenuous cytoplasmic strands.   If a cylindrical segment cut from a limb is implanted into the abdomen, the epidermal cells spread in the same fashion over the outer surface of the cuticle until they establish contact with cells migrating from the opposite end (Fig. 36).   Locke considers that ' contact ' means the continuous exchange of chemical stimulation with neighbouring cells; cells will migrate round a fragment of glass or plastic material and join up with cells on the other side.   But they will not migrate round a piece of millepore filter, even with the finest pores, presumably because the chemical factor can get through.[113]

IH E 2

Regeneration

The hypothesis outlined above, according to which a developing organule, such as a sensillum or a dermal gland, absorbs and unites with some inductor, or modifier, and by draining this from the surrounding zone inhibits the differentiation of a similar structure in its neighbourhood, can be extended so as to apply to the development of all the parts of the body.   The group of cells whose plasma has been determined in this way constitutes a morphogenetic ' field ' which is now committed

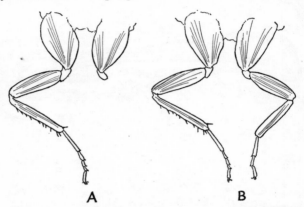

A   B

Fig. 37.   A, *Blattella* larva in which the metathoracic leg on the right was removed below the trochanter on the third day after hatching.   Moulting occurred at five days after hatching.   The operated limb shows only a smooth rounded papilla.   B, similar material in which the leg on the right was removed on the day of hatching.   Moulting occurred at 6 days after hatching.   The regenerated leg has only three tarsal segments and is a little smaller than the control (after O'Farrell and Stock).

to form the part in question.   Its parts become determined by the same process acting in a series of steps so that each subordinate part is committed to its future line of development and the rest is inhibited.

When the process is complete and an appendage such as a leg or a wing has been determined and developed up to say the pupal stage, the zone of cells around its base may still retain the capacity to repeat the entire process.   This capacity is normally inhibited by the presence of the appendage that is in course of growth.   But if the appendage is removed, this latent capacity is activated; renewed growth and differentiation take place and a new appendage is formed.[212]

In certain insects, notably in Phasmids and Blattids, the legs are very readily detached by ' autotomy ', usually at the level of the trochanter, and a new leg of normal form and size is regenerated. Regeneration can take place only during moulting and is, therefore, confined to the growing stages. We are here concerned with the interaction between the growth of the whole animal and the growth of the regenerating appendage; and particularly with the extent to which the endocrine system is involved.

It was noted by O'Farrell and Stock[216] that in the young cockroach *Blattella*, if a limb is removed late in the moulting cycle, just a day or two before the old skin is due to be cast, there is no delay in the moulting process and there is not even a partial regeneration of the missing limb —the wound is simply healed over and covered with cuticle (Fig. 37A). But if the limb is removed before the ' critical period ' when cell divisions should begin, regeneration takes place. Growth in the new limb is extremely active; but meanwhile growth in the rest of the body is held up; and not until regeneration is well advanced is general growth renewed. Thus moulting may be delayed for 10 days or so, and when it does take place the regenerated limb, which grows in folded form within the coxa of the old limb, is found to be almost equal in size and almost identical in organization with the normal limb on the other side (Fig. 37B). Often the size of the new instar is reduced so that the amount of growth from the one instar to the next is less than normal and the total number of moults before the adult stage is reached is increased. That happens in *Blattella*[216] and in *Leucophaea*.[216]

If the rudiment of one hind wing is extirpated in the last-stage larva of *Ephestia* and regeneration occurs, the growth of the opposite wing is arrested and emergence of the adult is delayed for some days. When the moth does emerge the regenerated wing is smaller than the control wing, and the control wing is somewhat smaller than usual.

Thus there is close integration between the growth of these regenerating appendages and the growth of the body as a whole. Whether the regeneration process inhibits other tissues from reacting to the hormone of the prothoracic gland, or whether the secretory activity of the prothoracic gland is inhibited during regeneration, is not known. In *Ephestia*, as studied by Pohley, the combined removal of brain, prothoracic glands and wings at an early stage in the last instar results in a permanent larva which regenerates the wing rudiments. Here regeneration seems to be independent of the moulting hormone.[216] Likewise in *Periplaneta*, the early stages of regeneration of the leg take place equally well in the absence of the central nervous system, but in the later stages the organization of the limb becomes defective; it is usually smaller, the musculature is very incomplete and it cannot, of course, be moved by the insect.[143, 156]

It has long been supposed that the growing tissues are communicating with one another and controlling one another by chemical means[211] (see p. 108). We had one example of this in the normal metamorphosis of the last larval stage in *Rhodnius*, where the epidermis of the abdomen halts its development while the complex outgrowths of the thorax and genitalia are being formed (p. 57).

From time to time the process of regeneration may go wrong. Removal of an antenna in *Carausius* may be followed by regeneration as a leg. This anomaly, where the appendage proper to another segment of the body is developed, is termed ' homoeösis '. It may occur

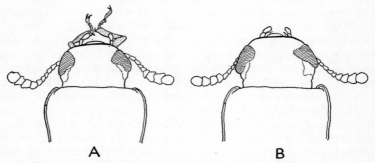

A   **B**

FIG. 38. A, ' labiopedia ' in *Tribolium* in which the labial palpi are replaced by perfect but diminutive prothoracic walking legs. B, normal adult for comparison (after Daly and Sokoloff).

regularly in certain genetic strains, as in ' aristopedia ' in *Drosophila* where the arista of the antenna may develop as a leg; or ' proboscipedia ' where the labella of the proboscis may likewise be replaced by legs (Fig. 38). Hadorn[70] has shown that fragments of imaginal discs in *Drosophila* can undergo ' transdetermination ' and some of the cells can become transformed into other cell types which are not characteristic of the original anlage. Cells of the third antennal segment and arista will give rise to tarsal structures. All such abnormalities can be pictured as resulting from the activation of new components of the gene system which in turn evoke characters that ought to have appeared elsewhere. But it must be pointed out once more that these chemical evocators and modifiers are at present purely hypothetical; the sort of effects ascribed to them closely resemble the known effects of the juvenile hormone, for example, but even the existence of such agents has not yet been proved.

Polymorphism within the species

The problem of differentiation concerns the regulation of ' polymorphism ' in the various parts of the body. In this section we are concerned with polymorphism in the members of the species. In many insects which occur naturally in more than one form the characters are controlled by genetic factors. The different forms exist side by side in the natural environment, the relative abundance of each form depending on the pressure of natural selection to which each is exposed. This phenomenon is called ' balanced polymorphism ' of which many examples are known.[98] There are other insects in which the different forms occur in individuals of constant genetic constitution. These forms likewise, of course, are produced by the action of genes; but the gene function is ' switched ', or latent genes are brought into action, under the impact of some change in the environment.[98]

In both these types of polymorphism chemical factors circulating in the blood may play a part. Most genes are said to be 'autonomous '— by which is meant that their action is confined to the cells in which such genes occur. Consequently, in ' mosaic ' individuals, in which different parts of the body have a different genetic constitution, the result is patchy: a mutant gene operating in some cells does not influence and is not influenced by the remainder of the body. If the two eyes in a transverse sexual mosaic (gynandromorph) of *Drosophila* carry the genes of different eye-colour mutants, neither is affected by the constitution of the other.

There is, however, one classic exception: ' vermilion ' eye-colour fails to develop in one eye of a mosaic if the other eye is of the wild type: its colour is influenced by the genetic constitution of other parts of the body. The circulating factor was at one time referred to as a ' gene hormone ', but this was not a very satisfactory term. The classic investigations by Ephrussi and Beadle on *Drosophila*[218] and of Kühn, Butenandt and others on *Ephestia*[218] (in which an homologous gene is responsible for the normal black eye colour) led to the biochemical explanation. The mutant insects are deficient in the enzyme which converts tryptophane into kynurenine which then goes by way of hydroxykynurenine to form the brown ommochrome pigment of the eyes. If kynurenine itself is injected into the mutant forms they can be induced to form normally pigmented eyes. This work is of some historic importance as being one of the first and clearest examples to support the ' one gene, one enzyme ' hypothesis which forms the basis of much modern ' molecular ' biology.

We saw that the arrest of hormone secretion which results in diapause is brought about by certain signals from the environment, notably the day-length (p. 31). At the same time, there often occurs a 'gene switch' that results in a striking change in structure or in pigmentation. In the butterfly *Araschnia levana* the seasonal forms are closely tied to diapause in this way (Fig. 39). Exposure of the larva to a 'long day'

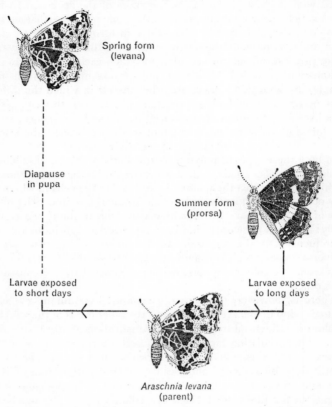

Spring form
(levana)

Diapause
in pupa

Summer form
(prorsa)

Larvae exposed
to short days

Larvae exposed
to long days

Araschnia levana
(parent)

FIG. 39. Control of diapause and the appearance of spring (levana) and summer (prorsa) forms of the butterfly *Araschnia levana* by regulating the photoperiod during the life of the caterpillar.

exceeding 16 hours is followed by a pupa which develops immediately without diapause and gives rise to the dark summer form *prorsa*. Larvae exposed to a short day of 8 hours give rise to pupae which pass through a prolonged diapause before they complete their development and emerge as the pale spring form *levana*.[98]

Other types of adversity in the environment may induce switches of this kind.   In the brown leafhopper *Nila parvata* which infests rice fields in Japan, the macropterous (long-winged) form appears under unfavourable conditions of nutrition, and enables the species to migrate by flight.   In their new habitat the migrants produce brachypterous (short-winged) forms for two generations, resulting commonly in an overpopulation which leads to the renewed appearance of the emigrating macropterous form.[100]

Interesting examples of nervous stimuli producing a genetic switch are seen in the pupae of certain butterflies (*Pieris*, *Vanessa*, etc.) which can become dark or pale according to the background on which they rest (Fig. 28); and in certain caterpillars, such as *Biston betularius*, which can become purplish-brown, smooth and shiny like a birch twig, or greenish-grey with the faintly granular colouring of a naked oak twig, or mottled,

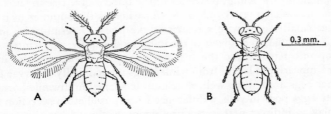

FIG. 40.   Dimorphism in the male *Trichogramma semblidis* (after Salt). A, winged male reared in egg of Lepidoptera; B, apterous male reared in egg of *Sialis*.

dark grey and white, like a twig covered with lichen.   Poulton, writing in 1903, recognized that these were genetic types arrived at by natural selection; but they are evoked by gene switching in accordance with the visual impressions received by the insect.[98]   Here again, hormonal agents are presumably concerned in controlling the changes.   Many grasshoppers respond in the same way.[218]

When larvae of the Chalcid egg parasite *Trichogramma semblidis* develop in the eggs of the alder fly *Sialis*, the resulting males are apterous; whereas when the larvae develop in the eggs of *Ephestia* and other Lepidoptera, the males are winged and show other striking differences in the form of the antennae and legs (Fig. 40).[212]   Perhaps this is an effect of nutrition; perhaps it results from differences in oxygen supply; but whatever the immediate cause the result is clearly due to a genetic switch comparable with that brought about by the juvenile hormone.

The effect of nutrition in producing switches of this kind is well seen in other Hymenoptera.   The queen honey-bee is of the same genetic constitution as the workers.   The distinctive changes in morphology are induced by nutrition; if the eggs are removed from the worker brood cells immediately after laying and placed in queen cells so that the larvae are fed continuously with ' royal jelly ', they develop into queens.   None of the characteristic compounds in royal jelly (biopterin, neopterin, pantothenic acid, etc.) will replace the active substance.   Royal jelly freeze-dried and stored under nitrogen at $-20°$ seems to retain its full activity; but dialysable factors are necessary.   It appears therefore that there is indeed a specific queen-producing material in royal jelly.[162, 208]   The juvenile hormone does not seem to be involved.

Likewise among ants, there is some evidence that abundant feeding on protein-rich food favours the development of queens.   But in *Formica* it seems that partial determination may occur in the preceding generation, for the winter ovarioles of the queens give rise to large eggs with an oösome (p. 111) rich in RNA; whereas the summer ovarioles produce smaller eggs with a restricted oösome.   Determination in the eggs is not final, but given a medium food supply the winter eggs develop into females and the summer eggs into workers, with ' inter-castes ' appearing if larvae from winter eggs are fed badly or larvae from summer eggs are fed too well.[102]   Since deficiency in juvenile hormone leads to eggs poor in yolk, a difference in corpus allatum secretion may well play an indirect part in the production of the worker caste.

Among termites, chemical substances produced by one form control the form of other individuals in the colony.   We shall consider this phenomenon in dealing with ' pheromones ' in Chapter 8.   But there is evidence that in the actual production of the soldier caste the juvenile hormone plays an important part; for worker larvae or nymphs receiving implants of extra corpora allata give rise to soldiers.[98]   Corpora allata taken from *Periplaneta* have the same effect.[109]

There has long been a belief that the change from the solitary non-migratory phase in locusts to the gregarious migratory phase might be brought about by hormonal differences.   There is some evidence that the solitary phase is a partially larval or ' neotenic ' form which still retains its thoracic glands in the adult stage and that these characters are due, at least in part, to increased production of juvenile hormone.[98, 216]   It has long been known that dense crowding of locusts in the solitary phase leads to a switch over to the ' gregarious ' phase in the course of a generation or so.   Crowding of mature females in *Locusta* leads to a reduction in the size and activity of the corpora allata and a decrease in the rate of egg laying.   On the other hand implantation of corpora allata into gregarious females increases their fecundity and leads them to produce offspring of more ' solitary ' characters: there is an increase in beige larvae and a decrease in black larvae.   Whereas extirpation of the corpora allata has the converse effect: it causes ' solitary ' females to produce young with phase characters of gregarious type.[31]

The green/brown polymorphism seems to be general throughout the Acridoidea, and excess of juvenile hormone following implantation of active corpora allata results in the green coloration.[172, 218]   The green

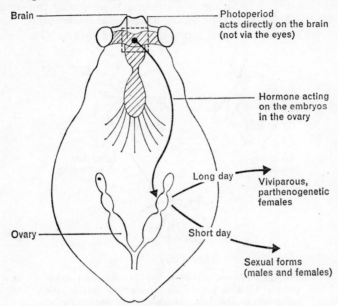

FIG. 41.   Control of reproductive polymorphism in the pea and bean aphid *Megoura* by regulating day length.

and brown ground colours correspond to different oxidation states of a protein-bound biliverdin.[154]

In the case of Aphids, such as the pea and bean Aphid *Megoura*, exclusively female parthogenetic forms appear all through the summer months when the day-length exceeds 14 hours. When the day-length is reduced below 14 hours, the embryos develop into males and females which reproduce sexually and lay overwintering eggs. By means of ingenious experiments in which the light beam was directed down a fine metal tube which could illuminate only a minute area of the aphid, it was shown by Lees that the photoperiod is perceived not by the eyes but directly by the brain, perhaps by neurosecretory cells in the pars intercerebralis (Fig. 41). Under the action of varying photo-period the brain apparently gives rise to differing hormonal stimuli which act upon the embryos in the ovary and so determine whether these develop solely into viviparous parthenogenetic females, or into sexual forms with both sexes present and with oviparous females.[110]

In most of the examples of polymorphic differentiation that have been mentioned, the nature of the hormonal or other stimuli that are immediately concerned in the switch over is not known. But in many Aphids that are reproducing by parthenogenesis during the summer months, the females may be either wingless neotenic forms, closely resembling larvae in outward appearance, or they may be winged forms

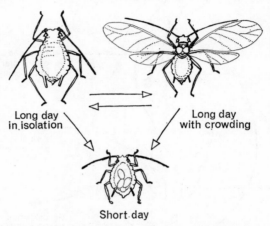

Long day
in isolation

Long day
with crowding

Short day

Fig. 42.   Polymorphism in Aphids.   Above, parthogenetic females producing live young during the
summer; to the left, wingless form appearing in
isolated individuals; to the right, winged form developing in crowded colonies.   Below, egg-laying
(bisexual) female appearing during the short days of
autumn.

with truly adult external characters (Fig. 42).   The wingless forms appear so long as the insects have plenty of space on the plants; they seem to result from an excessive production of juvenile hormone.   The winged forms appear as soon as there is any overcrowding among the Aphids and seem to be associated with a reduction in juvenile hormone secretion.[110]   If the first instar larvae of the cabbage aphid *Brevicoryne* are treated with materials with juvenile hormone activity at a suitable dose they show no morphological abnormality but a significant increase in the production of apterae.[210]

The environmental stimuli that induce the various morphological changes in Aphids, and perhaps the internal factors also, vary greatly in different species. A few examples may be quoted. In *Aphis cracivora* a very mild degree of crowding will cause the adult female to switch from producing apterous to alate progeny: brief controlled encounters between two adult aphids is sufficient. The stimulus is tactile. In the same species the condition of the host plant, acting either before or after birth, can influence the development of winged forms: plants in poor condition give more alates. High temperatures and long days both favour apterous development.[88] By rearing *Myzus persicae* on a wholly synthetic diet it could be shown by Mittler and Dadd[134] that even a change in a single amino acid in the diet can bring about changes in the number of winged forms produced. Thus, 94 per cent of more than 2,000 larvae reared on a diet containing isoleucine became alatae; compared with only 28 per cent on a diet lacking isoleucine.

Sexual dimorphism

The example of dimorphism that has been most closely studied is the difference between the sexes. Sex is determined primarily by the genetic constitution. That means that the genes are responsible for producing factors which determine maleness and femaleness, and that the balance of these sex determiners is such that if the genetic make-up is male, male characters are developed and *vice versa*. The genes in question are variously distributed in the sex chromosomes and in the autosomes. In mammals and in some Crustacea the genetic constitution in regard to sex can be over-ridden by the sex hormones: administration of steroid sex hormones to the female mammal during pregnancy results in a partial reversal of sex in the offspring; and the secondary sexual characters of mature birds can be reversed by castration or by the implantation of gonads.

In the insect the secondary sexual characters cannot be altered in this way. The genes controlling sex exert their action solely within the cells which carry them. If, therefore, the genetic constitution in respect to sex is different in different parts of the body, sexual mosaics or 'gynandromorphs' are produced. But if the sex determiners are incorrectly balanced, as when different races of the gypsy moth *Lymantria*[218] or the psychid moth *Solenobia*[218] are crossed, or in various species of hybrids, 'intersexes' may result. In the most extreme forms it is possible to have normally functioning males or females with the genetic constitution of the opposite sex. In the less extreme cases the reversal of sex is partial and affects some organs only.

Partial sex reversal may also be brought about by parasites. The classic example is the effect of *Stylops* on *Andrena* and other solitary bees, in which parasitized males acquire some of the characters of

IH F

females (chiefly pigment characters in the cuticle) and parasitized females come to resemble males. These changes seem to be the result of some kind of starvation effect by the parasite changing the balance of sex determiners.[218]

In the case of the midge *Chironomus* parasitized by the nematode *Mermis* the ovaries are destroyed in the last larval stage. The secondary sexual characters, in the fore-legs, antennae, etc., are not affected, but an unusual type of intersex is developed in which male internal and external organs of sex are found in an insect that is otherwise female. This suggests that in *Chironomus*, in addition to the balance between male-determining and female-determining factors, there may be a third ' epigenetic ' sex differentiating factor represented by a sex hormone controlling the development of the external genitalia.[218] Intersexual males of Chironomids can also be produced.[218] Indeed the peculiarities of the intersexes in *Chironomus* seems to depend on the uniformly ' coarse ' structure of the sexual mosaic: the intersexes with male sexual appendages are always cytologically normal males and those with female cerci are cytologically normal females.[218]

The sex determining mechanism may be influenced by temperature. In the normal strain of the hymenopterous parasite *Ooencyrtus submetallicus* the environmental temperature determines whether the progeny are male, female or intersexual.[226] In the familiar stick insect *Carausius morosus* all the insects are commonly females; but if the temperature of incubation for the eggs is raised to 30°C during the first third of embryonic life, they all develop into males.[218] And in the mosquito *Aëdes stimulans* and other species of the subgenus *Ochlerotatus* the normal sex ratio appears if the larvae are reared at a low temperature; but larvae reared in warm water show varying degrees of feminization extending up to almost complete females.[83] The effect of starvation on sex is seen in the hymenopterous parasite *Nasonia*. When the female has continuous access to the puparia of flies in which to lay its its eggs, the offspring consist of 25 per cent males. If they have access to puparia for only one hour per day the eggs undergo resorption and the percentage of males rises along with the increased number of eggs undergoing resorption at the time of oviposition.[99]

There is more direct evidence for a hormonal effect in the glow-worm *Lampyris noctiluca* where Naisse[137] has shown that the testes of the larva are the source of an androgenic hormone which controls all primary and secondary male sexual characters. This hormone, which is capable of masculinizing a female larva, is secreted in a mesodermic tissue, the apical tissue, present in all the testicular follicles. This tissue is characteristic of the testes of Lampyrids. The interpretation of these results is made difficult by the fact that the female *Lampyris* is a neotenic form which retains many larval characters; for example, the prothoracic glands which degenerate rapidly in the pupal stage of the male, persist to some extent up to the imaginal stage of the female. What is peculiar about *Lampyris* is the late occurrence of sexual differentiation. In most insects sex is clearly determined at eclosion from the egg; in *Lampyris* it is only at the fourth larval stage that testes and ovaries become distinguishable. When the apical tissue has become differentiated the implantation of the testis into a female larva causes total masculinization of both primary and secondary sexual characters.[137]

8: Pheromones

It has long been known that insects can influence other members of their species by means of chemical stimuli. The analogy between this mode of communication within an insect population, or an insect society, and the communication within the body by means of hormones, led to these substances being called ' ectohormones ' (Bethe) or sometimes ' social hormones '. In some cases, as we shall see, the substances involved do appear to have a hormone-like effect on growth, but as a rule they seem to act only on the sense organs. It was therefore proposed by Karlson and Lüscher[25] that a new term ' pheromone ' should be used to cover the wide range of substances secreted to the outside by the individual and received by other individuals of the same species in which they elicit some characteristic response.

These ' chemical messengers which act within the species ' have exceedingly diverse biological effects and they differ widely in their mode of action; but in practice the term ' pheromone ' has proved to be a very useful one in stimulating and coordinating work in this field. Renewed interest in the subject has coincided with the development of chemical procedures for separating and characterizing compounds available in very small amounts, with the result that the chemistry and mode of action of pheromones is a subject that is expanding very rapidly at the present time.[25]

Pheromones in aggregation, dispersal, trail formation and social coherence

Many insect species aggregate together in large numbers, and many factors are involved in this response; but odorous chemicals certainly play a part. Body lice *Pediculus*,[218] the cockroach *Blattella*,[84] and the Scolytid *Dendroctonus*[157] are attracted by the faeces of their own species. One could imagine how some incidental component of the faeces might come to be important as a stimulus to aggregation and later be specifically produced for this purpose.

Many insects produce ' defensive secretions ' which are repellent to potential predators. These comprise a vast range of saturated and unsaturated hydrocarbons and their derived alcohols and aldehydes (ranging from hexenal and tridecene to pentadecane), fatty acids (such as butyric and caproic acids), a great number of phenolic substances and quinones (such as phenol, m-cresol, benzoquinone, toluquinone, etc.) and a variety of terpenoids (such as geraniol, farnesol, citral, citronellol, etc.).[209]

Defensive secretions as such would not fall within the definition of the pheromones. But what are commonly regarded as repellent or defensive secretions against outsiders, can be utilized within the species for recognition, aggregation and trail formation. The trail pheromone of *Zootermopsis* contains many compounds, but one of the most active is caproic acid.[95] The Nassanoff glands of the honey-bee worker secrete a mixture of terpenoids (citral, geraniol, etc.) and this odour is used by foragers as an attractant to secure recruits for their expeditions to rich collecting grounds.[26] In *Lasius niger*[71] and in Ponerine ants generally, the contents of the stomach and hindgut will serve as an effective scent trail; in Myrmecine ants the secretion of the poison gland is used.[12]

Defensive secretions may also play the opposite role: they may serve to alert and disperse aggregations. That is seen in *Formica rufa* where formic acid is a potent means of defence and offence but serves also to alert the colony to a threatened attack and so to induce a general agitation throughout the nest. The citral produced by the mandibular glands of the leaf-cutting ant *Atta* likewise serves this same dual purpose of repellency and alarm; and in the cotton-stainer *Dysdercus*, which produces a mixture of hydrocarbons ranging from hexenal to pentadecane, this product can serve both as a repellent against predators and as a warning signal that causes dispersal of the aggregations which occur in this species.[28]

A very large number of ' alerting pheromones ' have been isolated and identified in ants and termites; many are terpenoids, others are aliphatic ketones and esters. In the fire ant *Solenopsis* a whole series of complex acts of behaviour, mass foraging, trail following, colony emigration, alarm, grooming and clustering are all largely induced by chemical releasers of this kind.[218] The mandibular glands of the queen honey-bee contain a variety of organic acids one of which, 9-hydroxy-decenoic acid, serves as a ' cohesion pheromone ' which keeps the swarm of worker bees clustered around their queen.[27] There is indeed evidence for the presence of widespread pheromones in other social insects, notably ants and termites, which are distributed throughout the colony by food sharing between adults and serve the purpose of colony cohesion, but the chemical nature of these materials is not known.

Among the social insects one of the best known chemical messengers of this kind is the ' queen substance ', 9-oxydecenoic acid, also secreted by the mandibular glands of the queen bee. This secretion is smeared over the integument of the queen during her grooming activities. The queen is usually surrounded by a circle of workers, all facing her and licking her and passing on the material in the course of food exchange with other workers. This is probably the chief means by which workers are kept informed of the presence of the queen. If she is removed a change in behaviour among the workers becomes apparent, and within 24 hours they begin building large ' queen cells ' to rear new queens. At the same time the ovaries of many of the workers ripen and they begin to lay eggs.

It may be that the absence of the same queen substance, in combination with a special queen scent, is responsible for both these effects.[25] The queen scent is a product of glandular cells in the mid-dorsal area of the abdomen[163] There is no evidence that queen substance or odour act directly on the reproductive system; it seems more probable that they act upon the sense organs and the central nervous system and thus cause indirectly a change in hormone secretion. During the first few days after removal of the queen the corpora allata of the worker bees increase in volume; their activity seems normally to be inhibited by the queen substance.[123]

Pheromones and mating

Pheromones play a large part in both sexes during the mating of insects; they serve both as attractants from a distance and as aphrodisiacs exciting the opposite sex to copulation. Sometimes both effects are given by the same substance. That is the case in the honey-bee in which the queen substance, 9-oxydecenoic acid, serves also as a sex attractant[218] which the drone will follow upwind, and as an excitant which is necessary to provoke copulation.[25] Sometimes the aphrodisiac secretion is most evident in the male. Many male Lepidoptera carry special scales (' androconia ') which are believed to produce an aphrodisiac secretion. The best evidence for this was obtained by Tinbergen[218] in *Eumenis semele* and other Satyrid butterflies, where the androconia of the male form an elongated patch across the centre of the fore-wing. These scent scales emit a scent which excites the female to accept the courting male. During an elaborate courtship, the knobs of the antennae of the female come to touch the patches of scent scales—then mating immediately follows.

If females of *Schistocerca* are reared without males several developmental defects become evident: the ovaries of some females fail to mature; or the rate of development is low and the percentage of resorbed oöcytes increases.[77] In the maturing male of *Schistocerca* carotinoids accumulate in the integument, giving the insect a bright yellow colour. This change is brought about by the corpora allata; if these glands are removed the colour does not develop and the operated males fail to accelerate maturation of females or of other males.[141] The normal mature male evidently produces a pheromone which accelerates maturation in both sexes.[141]

It has long been known that in cockroaches the dermal glands of both sexes give off sex pheromones which mix with the grease on the surface of the cuticle and act upon contact receptors on the antennae and mouth parts.[198, 218] In *Periplaneta* the sex attractant is produced principally by virgin females; mating depresses its production and it decreases with age.[218] In females of *Byrsotria* the corpus allatum controls sex pheromone production. The pheromone releases courtship behaviour in the male; he raises his wings, exposing the tergum, and apparently discharges a pheromone. The receptive female is attracted to the male and mounts and applies her mouth parts to his tergum.[169]

Among Lepidoptera there often seems to be a considerable overlap in the sexual scents of related species. Under laboratory conditions no specificity could be demonstrated in the attractive scents of female *Plodia interpunctella* and *Ephestia kuehniella*.[218] Even in the field, a certain number of *Lymantria monacha* males were attracted to *Lymantria dispar* females in spite of strong competition from wild *monacha* females.[218]

Scent organs in female Lepidoptera occur close to the sexual opening and take the form of tufts of modified scales or hairs with gland cells at their base, or a simple fold in the body wall consisting of a glandular epithelium covered by thin cuticle devoid of pores. Such organs are best developed among those Lepidoptera, such as Saturniidae and Bombycidae in which the eggs of the females are ripe for laying at the time of emergence from the pupa. In such species the males are attracted over large areas to the newly emerged females. The antennae in the male are not only highly developed and ' plumose ' in structure but the sense organs they carry respond specifically to the sex attractant, which is not masked even in the presence of strongly smelling substances.

In the case of the male silk moth *Bombyx mori* responding to 'bombycol', the name given to the female sex attractant, 50 per cent of the insects reacted by fluttering their wings in the presence of $10^{-4} \mu g$ of the pure substance, and significant responses were given with $10^{-5} \mu g$.

A calculation of the lowest concentration of the substance, in the air stream to which the antennae were exposed, that would give a response was 200 molecules per cm³ when the behaviour threshold was used as the test; 10,000,000 molecules per cm³ when the increase in nervous impulses, the ' electro-antennogram ' threshold, was used. At a concentration of 200 molecules per cm³, 40 out of 40,000 specialized receptor cells will receive one ' bombycol ' strike per second.[176]

The major part of the active material is deposited on the gland surface ready for dissipation; there is no large store. It is at a maximum at eclosion and production is suppressed after copulation. The behaviour of the female and the attractiveness of the gland secretion are not affected by allatectomy in the larval stage.[191] The chemical structure of ' bombycol ' has been elucidated and it has been synthesized;[23] it is *trans, cis*, 10, 12-hexadecadienol:

$$\overset{cis}{}\qquad\overset{trans}{}$$
$$CH_3(CH_2)_2CH\!\!=\!\!CH.CH\!\!=\!\!CH(CH_2)_8CH_2OH$$

The sex attractant obtained from the tip of the abdomen of the female gypsy moth *Lymantria dispar* has been used for many years in surveys aimed at discovering the presence of the species in a given locality. This attractant, named ' gyptol ', has likewise been isolated and synthesized by Jacobson *et al.*[86] It is similar to bombycol: *cis*, 10-acetoxy, 7-hexadecenol:

$$CH_3(CH_2)_5.\overset{\displaystyle |}{C}H\!-\!CH_2\ CH=CH(CH_2)_5CH_2OH.$$
$$\underset{\underset{\displaystyle O}{\parallel}}{O.C.CH_3}$$

The male wax moth *Galleria* produces a sex attractant to which the female responds; it is secreted by a gland that lies in a fold at the base of the fore-wing. Here the active principle is *n*-undecanal: $CH_3(CH_2)_9CHO$.[168] This belongs to the class of substances characteristic of repellent secretions in various insects and illustrates the specialized use of a common ingredient. This same phenomenon is perhaps to be seen also in the adult of the wireworm *Limonius* where the male beetles are attracted from a distance of 12m by the common fatty acid valeric acid produced by the female;[87] and in the mealworm *Tenebrio*, where the female sex pheromone which excites and attracts the male is in fact produced also by the male, but in smaller quantities.[203] In ants the different sexes and castes produce different odorous compounds, mixtures of terpenoids and indole bases. The mixtures are often qualitatively similar, but each species or caste produces a blend of distinctive proportions.[104]

Pheromones and the control of form

Most of the preceding examples of pheromones and their action have a somewhat tenuous connection with the subject of this book. But in termites pheromones play a major role in the control of caste production. It has long been recognized that all the eggs laid by the queen termite contain the genetic ' blueprints ' for all the castes characteristic of the species. What is uncertain is the method by which the diversion of development in a particular direction is brought about. The most widely held belief is that control is due to ' ectohormones ' (pheromones) liberated by each caste and tending to inhibit the development of the same caste; others believe that differences in nutrition are responsible; and yet others hold that a so-called ' group effect ' exists, and that the tactile and olfactory stimuli which result when the colony has a particular composition influence the type of development in the growing larvae. It may be that all these factors are concerned in varying degrees under different circumstances.

The example of ' ectohormonal ' control that has been most clearly demonstrated is in the European dry wood termite *Kalotermes*. The caste system in this insect is relatively simple (Fig. 43). A succession of larval stages ends in the full-grown larva or ' pseudergate '. The pseudergate may remain in that state indefinitely; it may proceed to grow into a winged adult (the primary reproductive form which after a mating flight will form a new colony); it may develop into a soldier; or it may turn into a secondary, or ' supplementary ', reproductive. In appearance the supplementary reproductives are intermediate between a larva and an adult that has shed its wings; they make their appearance in any colony which has lost the primary reproductive pair.

If the primary reproductives (the queen and her mate) are removed from a colony of *Kalotermes*, several supplementary reproductives develop within a week or ten days. As in the control of queen rearing by the queen substance of the honey-bee, this result does not happen if the queen is merely enclosed in a wire gauze chamber in such a way that the larvae can make contact with her; it does happen if she is enclosed by a double barrier of wire gauze. The pheromone responsible for preventing the development of supplementaries is excreted from the anus of the queen, taken up by the larvae and passed on to other members of the community. This same substance has another effect: if supplementary reproductives are already present it induces the pseudergates to destroy them. Moreover there are two pheromones that are

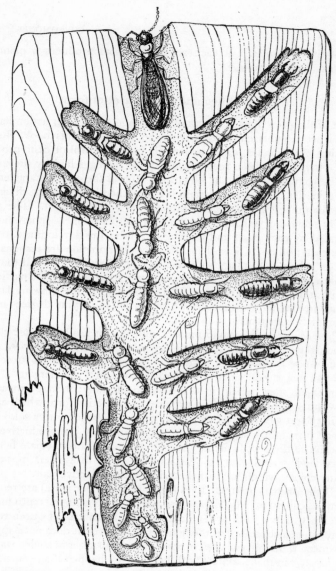

Fig. 43. Castes of *Kalotermes flavicollis*. After hatching from the egg the insect goes through about seven larval stages, then through a couple of 'nymphal' stages with visible wing lobes, which moult into winged adults. Any of these stages, from the 5th-stage larva to the 'nymph' may moult to a 'pre-soldier' and then to a soldier (right-hand side). Or any one of them may moult to produce a 'supplementary reproductive' (left-hand side). (After Lüscher).

more or less specific for the two sexes: the substances excreted by the male inhibits the formation of male supplementaries, and encourages the selective destruction of any male supplementaries already present. The product of the female has complementary effects. But there are differences in the controlling activity of the two sexes. A single reproductive female exerts, in the absence of a male reproductive, an almost complete inhibition on the transformation of female larvae and nymphs. Female reproductives thus always produce a sex-specific inhibitory pheromone, whereas male reproductives only produce a complete pheromone when they are in the presence of a female reproductive.[121]

In the same way the presence of soldiers inhibits the appearance of more soldiers, and if too many soldiers are present in the colony some of them are eliminated. It seems very likely that the soldiers also are producing other pheromones which prevent the development of new soldiers, but this has not yet been clearly proved.[98] We saw that the corpus allatum secretion can lead experimentally to the appearance of the soldier caste (p. 128). Indeed it appears that there is "not a single differentiation step in termites in which the corpora allata are not involved";[120] they seem to play a crucial part in caste differentiation. It may be that juvenile hormone is actually contained in the pheromones produced by the reproductives.[120]

References *

1. Bassurmanova, O. K., and Panov, A. A. *Gen. comp. Endocr.*, **9** (1967), 245–62 (Neurosecretion in *Bombyx*: fine structure and staining).
2. Baumann, G. *J. Insect Physiol.*, **14** (1968), 1459–76 (Juvenile hormone and cell membrane properties in *Galleria*).
3. Beck, S. D. *et al. Biol. Bull. mar. biol. Lab.*, *Woods Hole*, **126** (1964), 185–98; **128** (1965), 177–88; *J. Insect Physiol.*, **11** (1965), 297–303 ('Proctodone' and diapause in *Ostrinia*).
4. Becker, H. J. *Chromosoma*, **13** (1962), 341–84 ('Puffs' on salivary gland chromosomes of *Drosophila* and hormone action).
5. Bennet-Clark, H. C. *J. Insect Physiol.*, **8** (1962), 627–33 (Control of cuticle plasticity in *Rhodnius*).
6. Berendes H. D. *Chromosoma*, **22** (1967), 247–93 (Ecdysone and gene activity in *Drosophila*).
7. Berridge, M. J., and Patel, N. G. *Science*, **162** (1968), 462–3 (5-hydroxy-tryptamine stimulating salivary glands).
8. Berry, S. J., Krishnakumaran, A., Oberlander, H., and Schneider-man, H. A. *J. Insect Physiol.*, **13** (1967), 1511–37 (Hormones and injury controlling RNA synthesis in Saturniid moths).
9. Bessé, N. de, and Cazal, M. *C.r. hebd. Séanc. Acad. Sci.*, *Paris*, **266** (1968), 615–18 (Antidiuretic hormone from perisympathetic organs).
10. Bielenin, I. *Zoologica Pol.*, **13** (1963), 186–253 (Brain and endocrine glands in *Lecanium*).
11. Bloch, B., Thomsen, E., and Thomsen, M. *Z. Zellforsch. mikrosk. Anat.*, **70** (1966), 185–208 (Fine structure of neurosecretory cells in *Calliphora*).
*12. Blum, M. A. *A. Rev. Ent.*, **14** (1969), 57–80 (Alarm pheromones).
13. Bodenstein, D. *Biol. Bull. mar. biol. Lab.*, *Woods Hole*, **84** (1943), 34–58 (Hormones and growth of imaginal discs in *Drosophila*).
14. Bounhiol, J. J. *Bull. biol. Fr. Belg.*, Suppl. **24** (1938), 1–199 (Corpus allatum and metamorphosis in *Bombyx*).
15. Bowers, B., and Johnson, B. *Gen. comp. Endocr.*, **6** (1966), 125–9 (Distribution of neurosecretion along axons of Aphids).
16. Bowers, W. S., and Blickenstaff, C. L. *Science*, **154** (1966), 1673–4 (Hormones and ending of diapause in *Hypera*).
17. — *et al. Science*, **142** (1953), 1469–70; **154** (1966), 1020–1; *Life Sciences*, **4** (1965), 2323–31 (Chemicals with juvenile hormone activity).

* In the manuscript of this book almost every statement was documented by reference to original sources. For reasons of cost the list has been drastically reduced by the quotation of reviews where the original references can be found. Such reviews are marked with an asterisk (*). For the same reasons the titles of papers in journals are not given; instead an indication of the subject matter is given in parentheses.

18. BRADY, J. *J. exp. Biol.*, **47** (1967), 153–63 (Control of circadian activity rhythm in *Periplaneta*).

19. BROWN, B. E. *Gen. comp. Endocr.*, **5** (1965), 387–401 (Active substances from corpus cardiacum of *Periplaneta*).

20. — *Science*, **155** (1967), 595–7 (Neuromuscular transmitter in insect visceral muscle).

21. BÜCKMANN, D. *J. Insect Physiol.*, **3** (1959), 159–89 (Ecdysone and colour change in *Cerura* larva).

22. BURTT, E. T. *Proc. R. Soc. Lond.*, B, **126** (1938), 210–23 (Ring gland and growth in *Calliphora*).

23. BUTENANDT, A., and HECKER, E. *Angew. Chem.*, **73** (1961), 349–53 (' Bombycol ', sex attractant of *Bombyx*).

24. — and KARLSON, P. *Z. Naturf.*, **9b** (1954), 389–91 (Isolation of moulting hormone, ecdysone).

*25. BUTLER, C. G. *Biol. Rev.*, **42** (1967), 42–87 (Insect pheromones: review).

26. — and CALAM, D. H. *J. Insect Physiol.*, **15** (1969), 237–44 (Nassanoff gland pheromones in honey-bee).

27. — and SIMPSON, J. *Proc. R. ent. Soc. Lond.* (A), **42** (1967), 149–54 (Cohesion pheromone in honey-bee).

28. CALAM, D. H., and YOUDEOWEI, A. *J. Insect Physiol.*, **14** (1968), 1147–58 (Pheronomes in *Dysdercus*).

29. CARLSON, A. D. *J. Insect Physiol.*, **13** (1967), 1031–8; *J. exp. Biol.*, **48** (1968), 381–7 (Neurohumours and luminescence in *Photuris*).

30. CARLSON, J. G. *Chromosoma*, **5** (1952), 199–220 (Mitosis and differentiation of neuroblast in *Chortophaga*).

31. CASSIER, P. *Insectes soc.*, **12** (1965), 71–80; *Bull Soc. zool. Fr.*, **91** (1966), 125–48 (Hormonal control of phase characters in *Locusta*).

32. CAZAL, M., and GIRARDIE, D. *J. Insect Physiol.*, **14** (1968), 655–68 (Humoral control of water balance in *Locusta*).

33. CHASE, A. M. *Nature*, **215** (1967), 1516–17 (Metabolic poisons and adult development in *Tenebrio*).

34. CLARET, J. *Annls. Endocr.*, **27** (1966), 311–20 (Pigmentation of head capsule and photosensitivity in *Pieris*).

35. CLARKE, K. U., and GILLOT, C. *J. exp. Biol.*, **46** (1967), 13–34 (Metabolic effects of removal of frontal ganglion in *Locusta*).

36. — and LANGLEY, P. A. *Gen. comp. Endocr.*, **2** (1962), 625–6; *J. Insect Physiol.*, **9** (1963), 411–21 (Role of the frontal ganglion and growth in *Locusta*).

37. CLEVER, U. *Devl. Biol.*, **6** (1963), 73–98 (Ecdysone concentration and puffing pattern of salivary gland chromosomes in *Chironomus*).

38. COLES, G. C. *J. exp. Biol.*, **43** (1965), 425–31; *J. Insect Physiol.*, **11** (1965), 1317–30; **12** (1966), 1029–37 (Protein synthesis and yolk formation in *Rhodnius*).

*39. COLHOUN, E. H. *Adv. Insect Physiol.*, **1** (1963), 1–46 (Acetylcholine etc. in insects: review).

40. COOK, B. J., ERAKER, J., and ANDERSON, G. R. *J. Insect Physiol.*, **15** (1969), 445–55 (Action of amines on foregut of *Blaberus*).

*41. COTTRELL, C. B. *Adv. Insect Physiol.*, **2** (1964), 175–218 (Insect ecdysis and cuticular hardening and darkening: review).

*42. DANILYEVSKY, A. S. *Photoperiodism and seasonal development of insects,* Oliver and Boyd, Edinburgh, 1965.

*43. DAVEY, K. G. *Adv. Insect Physiol.*, **2** (1964), 219–45 (Control of visceral muscles in insects: review).

44. — *J. exp. Biol.*, **42** (1965), 373–8; *J. Insect Physiol.*, **13** (1967), 1629–36; **14** (1968), 1815–20 (Copulation and egg production in *Rhodnius*).

45. DAVIS, N. T. *J. Insect Physiol.* **10** (1964), 947–63; **11** (1965), 1199–1211 (Corpus allatum and reproduction in *Cimex*).

46. DOANE, W. W. *J. exp. Zool.*, **146** (1961), 275–98 (Effect of ripening of ovaries on fat body metabolism).

47. DOGRA, G. S. *J. Insect Physiol.*, **13** (1967), 1895–1906; *J. Morph.*, **121** (1967), 223–40 (Neurosecretory system in *Ranatra* and other Hemiptera).

48. DUKES, P. P., SEKERIS, C. E., and SCHMID, W. *Biochim. biophys. Acta*, **123** (1966), 126–33 (Action of ecdysone on isolated nuclei).

49. EDWARDS, J. S. *J. Insect Physiol.*, **12** (1966), 1423–33 (Neural control of metamorphosis in *Galleria*).

*50. ENGELMANN, F. *A. Rev. Entom.*, **13** (1968), 1–26 (Endocrine control of reproduction in insects: review).

*51. ETKIN, W., and GILBERT, L. I. (Eds.) *Metamorphosis*, Amsterdam, North Holland Publ. Co. 1969, 459 pp. (Four reviews deal with insect metamorphosis).

52. EWEN, A. B. *Can. J. Zool.*, **44** (1966), 719–27 (Corpus allatum and oöcyte maturation in *Adelphocoris*).

53. FINLAYSON, L. H., and OSBORNE, M. P. *J. Insect Physiol.*, **14** (1968), 1793–1801 (Peripheral neurosecretory cells in *Carausius* and *Phormia*).

54. — and WALTERS, V. A. *Nature*, **180** (1957), 713–14 (*Nosema* causing metathetely in Saturniids).

55. FLETCHER, B. S. *J. Insect Physiol.*, **15** (1969), 119–34 (Neurosecretory cell types in *Blaps*).

56. FRAENKEL, G., and HSIAO, C. *J. Insect Physiol.*, **11** (1965), 513–56 ('Bursicon' and tanning of insect cuticle).

57. FRONTALI, N., and NORBERG, K. A. *Acta physiol. scand.*, **66** (1966), 243–4 (Catecholamines in neurones of cockroach brain).

58. FUKUDA, S. *Annot. Zool. Japon.*, **35** (1962), 199–212 (Course of juvenile hormone secretion in *Bombyx*).

59. — and TAKEUCHI, S. *Embryologia*, **9** (1967), 333–53 (Diapause factor in suboesophageal ganglion of *Bombyx*).

*60. GABE, M. *Neurosecretion* (1966), 872 pp. Pergamon, Oxford.

61. GARCIA-BELLIDO, A. *Z. Naturf.*, **6** (1964), 492–5 (Male paragonia and fecundity of *Drosophila* females).

62. GELDIAY, S. *J. endocr.*, **37** (1967), 63–71 (Hormonal control of adult diapause in *Anacridium*).

63. GILBERT, L. I. *Comp. Biochem. Physiol.*, **21** (1967), 237–57 (Juvenile hormone and lipid metabolism in female *Leucophaea*).

*64. — *The Hormones*, Vol. 4 (1964), 67–134. Academic Press, New York (Hormones and insect growth: review).

65. — and GOODFELLOW, R. D. *Zool. Jb. Physiol.*, 71 (1965), 718–26 (Sterols and metamorphosis in *Hyalophora*).

66. — and SCHNEIDERMAN, H. A. *Trans. Am. microsc. Soc.*, 79 (1960), 38–67 (Bioassay for juvenile hormone).

67. GIRARDIE, A. *C.r. hebd. Séanc. Acad. Sci. Paris*, 261 (1965), 4876–8; *Bull. Soc. Zool. Fr.*, 91 (1966), 423–37 (Control of corpus allatum activity in adult *Locusta*).

68. GOSBEE, J. L., MILLIGAN, J. V., and SMALLMAN, B. N. *J. Insect Physiol.*, 14 (1968), 1785–92 (Nature of secretory neurones in brain of *Periplaneta*).

69. GRELLET, P. *C.r. hebd. Séanc. Acad. Sci. Paris*, 260 (1965), 5881–4 (Endocrine organs in embryo of *Scapsipedus*).

69a. GUPTA, B. L., and BERRIDGE, M. J. *J. Morph.*, 120 (1966), 23–81 (Neurosecretory nerves to rectum of *Calliphora*).

70. HADORN, E. *Devl. Biol.*, 13 (1966), 424–509 (Transdetermination in imaginal discs of *Drosophila*).

71. HANGARTNER, W., and BERNSTEIN, S. *Experientia*, 20 (1964), 392 (Scent trails in ants).

*72. HARVEY, W. R. *A. Rev. Ent.* 7 (1962), 57–80 (Metabolic aspects of insect diapause: review).

73. HASEGAWA, K., and YAMASHITA, O. *J. seric. Sci. Tokyo*, 36 (1967), 297–301 (Diapause hormone and ovary metabolism in *Bombyx* pupa).

*74. HERMAN, W. S. *Int. Rev. Cytol.*, 22 (1967), 269–347 (Prothoracic glands etc.: review).

75. — and GILBERT, L. I. *Nature*, 205 (1965), 926–7 (Neurosecretory cells in *Hyalophora*).

76. HIGHNAM, K. C., HILL, L., and GINGELL, D. J. *J. Zool.*, 147 (1965), 201–15 (Neurosecretion and water balance in *Schistocerca*).

77. — and LUSIS, O. *Quart. Jl Microsc. Sci.*, 103 (1962), 73–83 (Resorption of oöcytes in *Schistocerca*).

78. HILL, L. *J. Insect Physiol.*, 11 (1965), 1605–15 (Protein synthesis in fat body during yolk formation in *Schistocerca*).

79. — MORDUE, W., and HIGHNAM, K. C. *J. Insect Physiol.*, 12 (1966), 977–94; 1197–1208 (Frontal ganglion and oöcyte growth in *Schistocerca*).

80. HINKS, C. F. *Nature*, 214 (1967), 386–7 (Serotonin from neurosecretory cells in Lepidoptera).

81. HODGSON, E. S., and WRIGHT, A. M. *Gen. comp. Endocr.*, 3 (1963), 519–25 (Action of adrenaline etc. on insect nervous system).

82. HOLLINGSWORTH, M. J. *J. Morph.*, 115 (1964), 35–52 (Determination of chaetae in *Drosophila*).

83. HORSFALL, W. R., and ANDERSON, J. F. *J. exp. Zool.*, 156 (1964), 61–90 (Temperature and sex determination in mosquitoes).

84. ISHII, S., and KUWAHARA, Y. *Experientia*, 24 (1968), 88–9 (Aggregation in *Blattella*).

85. ISHIZAKI, H., and ICHIKAWA, M. *Biol. Bull. mar. biol. Lab.*, *Woods Hole*, **133** (1967), 355–68 (Brain hormone of *Bombyx*).

*86. JACOBSON, M., and BEROZA, M. *Science*, **140** (1963), 1367–73 (Chemical insect attractants: review).

87. — LILLY, C. E., and HARDING, C. *Science*, **159** (1968), 208–10 (Sex attractant in *Limonius*).

88. JOHNSON, B. *Entomologia exp. appl.*, **8** (1965), 49–64; **9** (1966), 213–22; 301–13 (Environmental control of wing polymorphism in *Aphis*).

89. — *J. Insect Physiol.*, **12** (1966), 645–52 (Fine structure of cardiac nerves in *Periplaneta*).

90. JOLY, L., PORTE, A., and GIRARDIE, A. *C.r. hebd. Séanc. Acad. Aci. Paris*, **265** (1967), 1633–5 (Fine structure of active and inactive corpus allatum in *Locusta*).

91. KARLINSKY, A. *C.r. hebd. Séanc. Acad. Sci. Paris*, **264** (1967), 1735–8; **265** (1967), 2040–2 (Corpus allatum and yolk formation in *Pieris*).

*92. KARLSON, P. *Pure & Appl. Chem.*, **14** (1967), 75–87 (Chemistry of insect hormones and pheromones: review).

93. — and BODE, C. *J. Insect Physiol.*, **15** (1969), 111–18 (Inactivation of ecdysone in *Calliphora*).

94. — and HOFFMEISTER, H. *Hoppe-Seyler's Z. physiol. Chem.*, **331** (1963), 298–300 (Conversion of cholesterol to ecdysone in *Calliphora*).

95. — LÜSCHER, M., and HUMMEL, H. *J. Insect Physiol.*, **14** (1969), 1763–71 (Chemistry of trail pheromone of *Zootermopsis*).

96. — and SEKERIS, C. E. *Acta Endocrinol.*, **53** (1966), 505–18 (Biochemical mechanisms of hormone action).

97. KATER, S. B. *Science*, **160** (1968), 765–7 (Release of heart accelerator from corpus cardiacum in *Periplaneta*).

*98. KENNEDY, J. S. (Ed.) *Symp. Roy. Ent. Soc. London* No. **1** (1961), 115 pp. *Insect Polymorphism* (Reviews by ten authors).

99. KING, P. E. *J. exp. Biol.*, **39** (1962), 161–5 (Egg resorption and sex ratio in *Nasonia*).

100. KISIMOTO, R. *Bull. Shikoku agric. Exp. Stn.*, **13** (1965), 1–106 (Polymorphism and population dynamics in *Nilaparvata*).

101. KOBAYASHI, M., and YAMASAKI, M. *Appl. Ent. Zool.*, **1** (1966), 53–60 (Sterol and protein as active components in extracts of *Bombyx* brain).

*102. KRAUSE, G., and SANDER, K. *Adv. Morphogenesis*, **2** (1962), 259–303 (Experimental embryology in insects: review).

103. LANGLEY, P. A. *J. Insect Physiol.*, **13** (1967), 1921–31 (Afferent stimuli via stomatogastric system controlling digestion in *Glossina*).

104. LAW, J. H., WILSON, E. O., and McCLOSKEY, J. A. *Science*, **149** (1965), 544–6 (Caste ' polymorphism ' in chemistry of ant pheromones).

105. LAWRENCE, P. A. *J. Cell Sci.*, **1** (1966), 475–98; *J. exp. Biol.*, **44** (1966), 607–20 (Gradients in determination in *Oncopeltus*).

106. — *J. exp. Biol.*, **44** (1966), 507–22; *Devl. Biol.*, **19** (1969), 12–40 (Hormonal control of development of hairs in *Oncopeltus*).

107. LEA, A. O. *J. Insect Physiol.*, **13** (1967), 419–29 (Neurosecretory cells and egg maturation in mosquitoes).

108. LEAHY, M. G. *J. Insect Physiol.*, **13** (1967), 1283–92 (Non-specificity of the paragonia factor in Diptera).

109. LEBRUN, D. *Annls. Soc. ent. Fr.*, **3** (1967), 867–71 (Corpora allata and polymorphism in *Calotermes*).

*110. LEES, A. D. *Adv. Insect Physiol.*, **3** (1966), 207–77 (Control of polymorphism in Aphids: review).

*111. — *The physiology of diapause in arthropods* (1955), 151 pp. Cambridge University Press.

112. LINZEN, B. *Naturwissenschaften*, **54** (1967), 259–67 (Biochemistry of ommochromes).

113. LOCKE, M. *J. Insect Physiol.*, **12** (1966), 389–95 (Cell interactions in wound healing in *Rhodnius*).

*114. — *Adv. Morphogenesis*, **6** (1967), 33–88 (Patterns and gradients in the integument of insects: review).

115. LOCKSHIN, R. A. *Science*, **154** (1966), 775–6 (RNA formation in early embryogenesis).

116. — and WILLIAMS, C. M. *J. Insect Physiol.*, **10** (1964), 683–9 (Hormonal control of muscle breakdown at metamorphosis in Saturniids).

117. LOHER, W. *Zool. Jb Physiol.*, **71** (1965), 677–84 (Hormonal control of oöcyte development in *Gomphocerus*).

118. — *Z. vergl. Physiol.*, **53** (1966), 277–316 (Control of egg ripening and sexual behaviour in *Gomphocerus*).

119. LÜÖND, H. *Devl. Biol.*, **3** (1961), 615–56 (Pattern formation in fragmented primordia of *Drosophila*).

120. LÜSCHER, M. *Proc. XVI Int. Congr. Zool. Washington*, **4** (1963), 244–50 (Functions of corpus allatum in termite development).

121. — *Insectes soc.*, **11** (1964), 80–90 (Production of supplementary sexual forms in *Kalotermes*).

122. — *J. Insect Physiol.*, **14** (1968), 499–511; 685–88 (Hormonal regulation of egg formation in *Nauphoeta*).

123. — and WALKER, I. *Rev. suisse Zool.*, **70** (1963), 305–11 (Mode of action of ' queen substance ' in honey-bee).

124. MADDRELL, S. H. P. *J. exp. Biol.*, **40** (1963), 247–56; **41** (1964), 163–76; 459–72; **45** (1966), 499–508 (Diuretic hormone in *Rhodnius*).

125. —*J. exp. Biol.*, **44** (1966), 59–68 (Control of mechanical properties of *Rhodnius* cuticle).

126. — PILCHER, D. E. M., and GARDINER, B. O. C. *Nature*, **222** (1969), 784–5 (Diuretic action of 5-hydroxytryptamine).

127. MANNING, A. *Nature*, **211** (1966), 1321–2 (Corpus allatum and sexual receptivity in *Drosophila*).

128. MEAD-BRIGGS, A. R. *J. exp. Biol.*, **41** (1964), 371–402 (Influence of the host on reproduction in the rabbit flea *Spilopsyllus*).

129. MERLE, J. *J. Insect Physiol.*, **14** (1968), 1159–68 (Effects of male paragonia on egg laying of female *Drosophila*).

130. MEYER, A. D., SCHNEIDERMAN, H. A., HANZMANN, E., and KO, J. H. *Proc. Nat. Acad. Sci.*, **60** (1968), 853–60 (Second juvenile hormone in *Hyalophora*).

131. MILLER, T. *J. Insect Physiol.*, **14** (1968), 1713–17 (Response of cardiac neurones of *Periplaneta* to acetylcholine).

132. — *et al. J. Insect Physiol.*, **14** (1968), 1099–1104; 1265–75 (Neurosecretory axons to heart of *Periplaneta*).

133. MILLS, R. R. *et al. J. Insect Physiol.*, **11** (1965), 1047–53; 1269–75; **12** (1966), 275–80; 1395–1401; **13** (1967), 815–20 (Tanning of cuticle in *Periplaneta*).

133a. MINKS, A. K. *Arch. Néerl. Zool.*, **18** (1967), 175–258 (Biochemistry of juvenile hormone action in adult *Locusta*).

134. MITTLER, T. E., and DADD, R. H. *Ann. Ent. Soc. Am.*, **59** (1966), 1162–6 (Nutrition and wing development in Aphids).

135. MORDUE, W. *J. Insect Physiol.*, **11** (1965), 493–511; 617–29; *Gen. comp. Endocr.*, **9** (1967), 406–17 (Corpus allatum and egg development in *Tenebrio*).

136. MUELLER, N. S. *Devl. Biol.*, **8** (1963), 222–40 (Moulting in embryo of *Melanoplus*).

137. NAISSE, J. *Archs. Biol. Liège*, **77** (1966), 139–201; *Gen. comp. Endocr.*, **7** (1966), 85–110 (Hormonal control of sex in *Lampyris*).

138. NATALIZI, G. M., and FRONTALI, N. *J. Insect Physiol.*, **12** (1966), 1279–87 (Purification of peptide hormones from corpus cardiacum).

139. NISHIITSUTSUGI-UWO, J., PETROPULOS, S. F., and PITTENDRIGH, C. S. *Biol. Bull. mar. biol. Lab., Woods Hole*, **133** (1967), 679–96 (Control of circadian activity rhythm of cockroach).

140. NORMANN, T. C. *Z. Zellforsch. mikrosk. Anat.*, **67** (1965), 461–501 (Neurosecretory system of adult *Calliphora*).

141. NORRIS, M. J. *et al. Nature*, **208** (1965), 1122; **219** (1968), 865–6 (Hormonal and social factors in sexual maturation of *Schistocerca*).

142. NOWOSEILSKI, J. W., and PATTON, R. L. *J. Insect Physiol.*, **9** (1963), 401–10 (Activity rhythm in *Gryllus*).

143. NÜESH, H. *A. Rev. Ent.*, **13** (1968), 27–43 (Nervous system in insect morphogenesis and regeneration).

144. NÚÑEZ, J. A. *Z. vergl. Physiol.*, **38** (1956), 341–54 (Control of water balance in *Anisotarsus*).

145. OBERLANDER, H., BERRY, S. J., KRISHNAKUMARAN, A., and SCHNEIDERMAN, H. A. *J. exp. Zool.*, **159** (1965), 15–28 (Nucleic acid synthesis in Saturniid moths under action of ecdysone).

146. ODHIAMBO, T. R. *Nature*, **207** (1965), 1314–15; *J. exp. Biol.*, **45** (1966), 45–61 (Metabolic effects of corpus allatum in *Schistocerca*).

147. — *J. Insect Physiol.*, **12** (1966), 995–1002 (Ultrastructure of corpus allatum in adult male *Schistocerca*).

148. OHTAKI, T. *Annotnes. zool. jap.*, **33** (1960), 97–103 (Hormone and pupal coloration in *Pieris*).

149. — *Japan J. med. Sci. Biol.*, **19** (1966), 97–104 (Arrest of pupation in *Sarcophaga*).

150. — MILKMAN, R. D., and WILLIAMS, C. M. *Biol. Bull. mar. biol. Lab., Woods Hole*, **135** (1968), 322–34 (Quantitative study of secretion and utilization of ecdysone in *Sarcophaga*).

151. OKASHA, A. Y. K. *J. exp. Biol.*, **48** (1968), 455–86; *J. Insect Physiol.*, **14** (1968), 1621–34 (High temperature and arrest of growth in *Rhodnius*).

152. ORR, C. W. M. *J. Insect Physiol.*, **10** (1964), 53–64; 103–19 (Hormones and nutrition in ovary development in *Phormia*).

153. OVERTON, J., and RAAB, M. *Devl. Biol.*, **15** (1967), 271–87 (Development of centrifuged eggs of *Chironomus*).

154. PASSAMA-VUILLAUME, M. *Bull. Soc. zool. Fr.*, **90** (1965), 485–91 (Green and brown pigment in *Locusta* and *Mantis*).

155. PASTEELS, J. M. *C.r. hebd. Séanc. Acad. Sci. Paris*, **261** (1965), 3884–6 (Neurosecretion in suboesophageal ganglion of *Carausius*).

156. PENZLIN, H. *Wilhelm Roux Arch. EntwMech. Org.* **155** (1964), 152–61 (Nervous system and regeneration in insects).

157. PITMAN, C. B., VITÉ, J. P., KINZER, G. W., and FENTIMAN, A. F. *J. Insect Physiol.*, **15** (1969), 363–6 (Aggregation pheromones in *Dendroctonus*).

158. RAABE, M. *Annls Endocr.*, **25** (1964), 107–12; *C.r. hebd. Séanc. Acad. Sci., Paris*, **263** (1966), 408–11 (Neurosecretory cells and colour change).

159. — *Bull. Soc. zool. Fr.*, **90** (1965), 631–54; *C.r. hebd. Séanc. Acad. Sci., Paris*, **262** (1966), 303–6 (Neurosecretion in ventral ganglia of Phasmids).

160. — *et al. C.r. hebd. Séanc. Acad. Sci., Paris*, **261** (1965), 4240–3; **263** (1966), 2002–5; **264** (1967), 77–80 (Perisympathetic organs in Phasmids).

161. RALPH, C. L. *J. Insect Physiol.*, **8** (1962), 431–9 (Heart accelerators etc. from corpus cardiacum of *Periplaneta*).

162. REMBOLD, H. *Umschau*, **14** (1967), 454–5 (Queen determining principle in ' royal jelly ').

163. RENNER, M., and BAUMANN, M. *Naturwissenschaften*, **51** (1964), 68–9 (Epidermal scent glands of queen bee).

164. RING, R. A. *J. exp. Biol.*, **46** (1967), 117–36 (Induction of diapause in *Lucilia* larvae).

165. RINTERKNECHT, E. *Bull. Soc. zool. Fr.*, **89** (1964), 383–92; **91** (1966), 645–54; 789–800 (Wound healing in *Locusta*).

166. RÖLLER, H. *Naturwissenschaften*, **49** (1962), 524 (Corpus allatum and egg production in *Galleria*).

167. — *et al. J. Insect Physiol.*, **11** (1965), 1185–97; **15** (1969), 379–89; *Recent Progress in Hormone Research*, **24** (1968), 651–80 (Chemistry of juvenile hormone).

168. — *et al. Acta ent. bohemoslov.*, **65** (1968), 208–11 (Sex attractant of *Galleria*).

169. ROTH, L. M. *et al. J. Insect Physiol.*, **10** (1964), 915–45; 965–75; **12** (1966), 255–65 (Sex pheromones in cockroaches).

170. ROTHSCHILD, M., and FORD, B. *Nature*, **201** (1964), 103–4 (Egg production in *Spilopsyllus* and reproductive hormones of the host).

171. ROUSSEL, J. P. *Bull. Sci. zool. Fr.*, **91** (1966), 379–91 (Role of frontal ganglion in insects).

172. ROWELL, C. H. F. *J. Insect Physiol.*, **13** (1967), 1401–12 (Corpus allatum and coloration in grasshoppers).

173. SAUNDERS, D. S. *J. exp. Biol.*, **42** (1965), 495–508; *J. Insect Physiol.*, **12** (1966), 569–81; 899–908 (Maternal induction of diapause in *Nasonia*).

174. SCHARRER, B. *Z. Zellforsch. mikrosk. Anat.*, **62** (1964), 125–48; **64** (1964), 301–26; **89** (1968), 1–16; **95** (1969), 177–86; *Arch. Anat. micr.*, **54** (1965), 331–42 (Fine structure of insect endocrine glands).

175. SCHEUER, R., and LÜSCHER, M. *Rev. suisse Zool.*, **73** (1966), 511–16 (Phases in ovarian competence in *Leucophaea*).

176. SCHNEIDER, D. *et al. Symp. Soc. Exp. Biol.* **20** (1965), 273–97; *Cold Spring Harb. Symp. quant. Biol.*, **30** (1965), 263–80; *Z. vergl. Physiol.*, **54** (1967), 192–209.

177. SCHREINER, B. *Gen. comp. Endocr.*, **6** (1966), 388–400 (Histochemistry of neurosecretory material in *Oncopeltus*).

178. SEHNAL, F. *J. Insect Physiol.*, **14** (1968), 73–85 (Corpus allatum and development of internal organs in *Galleria*).

179. SHAAYA, E., and KARLSON, P. *Devl. Biol.*, **11** (1965), 424–32 (Ecdysone in adult Lepidoptera).

180. SIEW, Y. C. *J. Insect Physiol.*, **11** (1965), 1–10; 463–79; 973–81 (Endocrine control of adult diapause in *Galeruca*).

181. SLÁMA, K. *J. Insect Physiol.*, **10** (1964), 283–303; 773–82; **11** (1965), 113–22 (Effect of hormones on growth, reproduction and metabolism in *Pyrrhocoris*).

182. — and HRUBEŠOVÁ, H. *Zool. Jb Physiol.*, **70** (1963), 291–300 (Effect of corpus allatum of *Pyrrhocoris* on metamorphosis and reproduction).

183. — and WILLIAMS, C. M. *Biol. Bull. mar. biol. Lab.*, *Woods Hole*, **130** (1966), 235–46 ('Paper factor' and metamorphosis in *Pyrrhocoris*).

*184. SMALLMAN, B. N., and MANSINGH, A. *A. Rev. Ent.*, **14** (1969), 387–408 (Cholinergic system in insect development: review).

185. SMITH, U., and SMITH, D. S. *J. Cell Sci.*, **1** (1966), 59–66 (Neurosecretion in corpus cardiacum of *Carausius*).

186. STAAL, G. B. *Koninkl. Nederl. Akad. Wetensch.*, **70** (1967), 409–18 (Plants as source of insect hormones).

187. STAY, B., and GELPERIN, A. *J. Insect Physiol.*, **12** (1966), 1218–26 (Control of oviposition in *Pycnoscelus*).

188. STEELE, J. E. *Nature*, **192** (1961), 680–1; *Gen. comp. Endocr.*, **3** (1963), 46–52 (Hyperglycaemic hormone in *Periplaneta*).

189. STEGWEE, D., KIMMEL, E. C., DE BOER, J. S., and HENSTRA, S. *J. Cell Biol.*, **19** (1963), 519–27 (Corpus allatum and reversible degeneration of flight muscle in *Lepinotarsa*).

190. — *et al. J. Insect Physiol.*, **8** (1962), 117–26; **10** (1964), 97–102 (Corpus allatum and oxidative metabolism).

191. STEINBRECHT, R. A. *Z. vergl. Physiol.*, **48** (1964), 342–56 (Sexual scent gland of *Bombyx*).

192. STOFFOLANO, J. G., and MATTHYSSE, J. G. *Ann. ent. Soc. Am.*, **60** (1967), 1242–6 (Diapause in *Musca autumnalis*).

193. STRAMBI, A. *C.r. hebd. Séanc. Acad. Sci.*, *Paris*, **264** (1967), 2031–4 (Neurosecretion and reproductive arrest in *Polistes*).

194. STRONG, L. *J. Insect Physiol.*, **11** (1965), 135–46; 271–80 (Endocrine organs and oöcyte growth in *Schistocerca*).

195. — *Nature*, **210** (1966), 330–1 (Frontal ganglion and egg development in *Locusta*).

196. — *J. exp. Biol.*, **48** (1968), 625–30; *J. Insect Physiol.*, **14** (1968), 1685–92 (Corpus allatum and fat body metabolism in *Locusta*).

197. STUMPF, H. F. *J. exp. Biol.*, **49** (1968), 49–60 (Epidermal gradients and differentiation in *Galleria* pupa).

198. STÜRCHOW, B., and BODENSTEIN, W. G. *Experientia*, **22** (1967), 851–3 (Sex pheromone in *Periplaneta*).

199. THOMAS, A. *C.r. hebd. Séanc. Acad. Sci.*, *Paris*, **267** (1968), 518–21 (Hormones and oviposition in *Carausius*).

200. THOMSEN, E. *et al. J. exp. Biol.*, **40** (1963), 301–21; *Z. Zellforsch. mikrosk. Anat.*, **75** (1966), 281–300 (Neurosecretion and production of mid-gut enzymes in *Calliphora*).

*201. TREHERNE, J. E. *The neurochemistry of arthropods*, Cambridge University Press, 1966, 154 pp.

202. — and SMITH, D. S. *J. exp. Biol.*, **43** (1965), 13–21; 441–54 (Acetylcholinesterase in central nervous system of *Periplaneta*).

203. TSCHINKEL, W., WILLSON, C., and BERN, H. A. *J. exp. Zool.*, **164** (1967), 81–6 (Sex pheromone in *Tenebrio*).

204. USHERWOOD, P. N. R. *et al. Nature*, **219** (1968), 1169–72; *J. exp. Biol.*, **49** (1968), 341–61 (Glutamate as neuromuscular transmitter in *Schistocerca*).

205. VROMAN, H. E., KAPLANIS, J. N., and ROBBINS, W. E. *J. Insect Physiol.*, **11** (1965), 897–904 (Corpus allatum and lipid turnover in *Periplaneta*).

206. WALL, B. J. *et al. Am. Zool.*, **5** (1965), 211; *J. Insect Physiol.*, **13** (1967), 565–78 (Control of water balance in *Periplaneta*).

207. WATSON, J. A. L. *J. Insect Physiol.*, **10** (1964), 305–17; 399–408 (Moulting and reproduction in *Thermobia*).

*208. WEAVER, N. *A. Rev. Ent.*, **11** (1966), 79–102 (Physiology of caste determination; review).

*209. WEATHERSTON, J. *Q. Rev.*, **21** (1967), 287–313; *Residue Rev.*, (1969) (Chemistry of defensive secretions in insects: review).

210. WHITE, D. F. *J. Insect Physiol.*, **14** (1968), 901–12 (Juvenile hormone and aptera production in Aphids).

211. WIGGLESWORTH, V. B. *Q. Jl Microsc. Sci.*, **77** (1934), 191–222; **79** (1936), 91–121; *J. exp. Biol.*, **17** (1940), 201–22 (Hormones controlling moulting and metamorphosis in *Rhodnius*).

*212. — *The physiology of insect metamorphosis*, Cambridge University Press, 1954, 152 pp.

213. — *J. exp. Biol.*, **32** (1955), 485–91 (Control of breakdown of prothoracic gland in *Rhodnius*).

214. — *J. exp. Biol.*, **32** (1955), 649–63 (High temperature and arrested growth in *Rhodnius*; quantitative requirements for ecdysone).

215. — *Symp. Soc. Exp. Biol.*, **11** (1957), 204–27; *J. Exp. Biol.*, **40** (1963) 231–45 (Mode of action of growth hormones in insects).

*216. — *Adv. Insect Physiol.*, **2** (1964), 247–336 (Hormonal control of growth and reproduction in insects: review).

217. — *Symp. Soc. Exp. Biol.*, **18** (1964), 265–81 (Homeostasis in insect growth).

*218. — *The principles of insect physiology*, 6th Edn. Methuen, London, 1965, 741 pp.

219. — *J. Insect Physiol.*, **15** (1969), 73–94 (Chemical structure and juvenile hormone activity).

220. — (1969) (unpublished observations).

*221. WILDE, J. de *A. Rev. Ent.*, **7** (1962), 1–26 (Photoperiodism in insects: review).

222. — *et al. J. Insect Physiol.*, **3** (1959), 75–85; **6** (1961), 152–61; *Konink. Nederl. Akad. Wetensch.*, (C) **71** (1968), 321–6 (Photoperiod controlling juvenile hormone secretion in *Leptinotarsa*).

223. — and FERKET, P. *Med. Rijksfacult. Landbouw. Wetensch.* Gent **32** (1967), 387–92 (Aging host plant controlling diapause in *Leptinotarsa*).

224. WILLIAMS, C. M. *Biol. Bull. mar. biol. Lab.*, *Woods Hole*, **134** (1968), 344–55 (Assay of ecdysone analogues on *Samia* pupae).

225. — *et al. Biol. Bull. mar. biol. Lab.*, *Woods Hole*, **127** (1964), 511–25; **128** (1965), 497–507 (Photoperiod and pupal diapause in *Antheraea*).

226. WILSON, F. *Aust. J. Zool.* **10** (1962), 349–59 (Temperature and sex determination in *Ooencyrtus*).

227. WYATT, G. R., and LINZEN, B. *Biochim. Biophys. Acta*, **103** (1965), 588–600 (Metabolism after injury in diapausing pupa of *Hyalophora*).

228. YAMASHITA, O., and HASEGAWA, K. *J. Sericult. Sci.*, *Tokyo*, **33** (1964), 115–23; 407–16; *J. Insect Physiol.*, **12** (1966), 325–30; 957–62 (Mode of action of diapause hormone in *Bombyx*).

229. ZDÁREK, J., and SLÁMA, K. *J. Insect Physiol.*, **14** (1968), 563–7 (Mating behaviour of adultoid larvae in *Pyrrhocoris*).

230. ZWICKY, K., and WIGGLESWORTH, V. B. *Proc. R. ent. Soc. Lond.* (A) **31** (1956), 153–60 (Oxygen consumption during moulting cycle in *Rhodnius*).

Index

Acetylcholine, 101

Acheta: activity rhythm, 96

Acronycta: photoperiod and diapause, 32, 34

Activation, 16

Activation hormone, 5, 13, 23

Adelphocoris: corpus allatum and brain in reproduction, 78

Adrenalin, 101

Aëdes: arrest of growth by lack of oxygen and salts; dormancy in mature embryo, 38; corpus allatum and reproduction, 73; neurosecretory cells and egg production, 78; humoral factor from spermathecae and ovulation, 83; male accessory gland secretion as oviposition stimulus, 84; temperature and sex, 132

Aleurodidae:desiccation and dormancy of egg, 37

Υ-Aminobutyric acid (GABA), 102

Anacridium: neurosecretion and adult diapause, 86

Anax: photoperiod and diapause, 34

Androconia, 136

Anisolabis: moulting of adult, 12; reversal of metamorphosis, 58; corpus allatum and reproduction, 73

Anopheles: gonadotrophic dissociation, 86

Antheraea: effect of activation hormone, 15; temperature and diapause development, 35; photoperiod ending diapause, 36

Apanteles: oösome, 111

Aphididae: neurosecretory axons in sympathetic system, 93; mobilization of flight muscle protein, 119; juvenile hormone and aptera formation, 130; crowding and alate production, 131

Aphrodisiacs, 136

Apis: omission of pupal stage, 63; caste determination, 127; Nassanoff gland and pheromone production,

135; cohesion pheromone, 135; queen substance, as sex attractant, 136

Araschnia: seasonal dimorphism, 126

Arctias: brain removal and diapause, 29

Aristopedia, 124

Atta: citral as alarm pheromone, 135

Austroicetes: egg diapause, 35

Autotomy, 123

Barathra: photoperiod and diapause, 34; temperature and diapause development, 36

Berlese theory, 42

Biston: prolonged diapause, 36; colour change in response to background, 127

Blaberus: discharges of neurosecretion after electrical stimulus to brain, 82; conduction of impulses by neurosecretory neurons, 93

Blattella: 'group-effect' on growth, 31; oötheca as inhibitory stimulus to corpus allatum, 80; limb regeneration, 123; aggregation by faeces, 134

Bombycol, 137

Bombyx: castration, 1; prothoracic gland, 4; neurosecretory cells, 7; activation hormone and corpus allatum, 15; nature of activation hormone, 23; isolation of ecdysone, 24; ecdysone in adult, 27; diapause in egg, 38; voltinism, 38; diapause hormone, 39; corpus allatum and metamorphosis, 49; secretion of juvenile hormone by isolated corpus allatum, 52; egg ripening without corpus allatum, 75; neurosecretion and oviposition, 83; hypoglycaemic effect of diapause hormone, 109; female sex attractant: 'bombycol', 137

Brevicoryne: juvenile hormone and aptera production, 130

153